Management of
Krishi Vigyan Kendras

Management of

Krishi Vigyan Kendras

Nikulsinh M. Chauhan

BIOTECH

2013

BIOTECH BOOKS®

ISBN 978-81-7622-282-2

Published by: **BIOTECH BOOKS®**
4762-63/23, Ansari Road, Darya Ganj,
New Delhi - 110 002
Phone: +91-011-23262132
E-mail: biotechbooks@yahoo.co.in

Printed at: **Chawla Offset Printers**
Delhi - 110 052

PRINTED IN INDIA

Dedicated to

my

Beloved

Late Brother

Dilipsinhji Chauhan

Preface

This book will be useful to agricultural researchers, educationists and extension workers to run on *information super highway with full of speed and impact oriented TOT*. India's population has largely crossed 1 billion and by 1035 experts project that India will overtake China as the world's most populous country. At the same time India already holds first place in another category; the county is home to the world's largest number of poor people. Today estimated 350 million out of 1 billion Indians fall between the poverty line. This number represents a grim reality; while India claims self-sufficiency in production of grains, two-third of the Indian children under the age of 5 are malnourished. Infant mortality is 65 per 1000 live births, compared with 33 for China. Adult literacy rates of 65 per cent for men and 38 per cent for women fall far short of those in China 90 per cent for men and above 73 per cent for women.

Can India achieve its own economic miracle, pushing back poverty and stimulating a vibrant economy with opportunities for its entire people? The task for India's policymakers is to recharge agricultural growth, enabling poor people to contribute their skills and energies to the country's economic activities to get the enhanced benefits of economic success.

Indian agricultural sector is vast and has continued to be the backbone of our economy. It supports 70 per cent of the country's population, which contributes 30-35 per cent of GDP and generates about 20 per cent of export earnings (Mehta, 1998). It provides employment for approximately 62 per cent of the workforce (Singh, 2000). It is indeed heartening that during the 55th year of India's independence, we could reap a record harvest of 208 million tonnes of food grains (Anonymous, 2002).

During the past four decades, spectacular progress has been achieved in agricultural production in the country, which has been well chronicled and universally recognized. In the oil seed sector, there has been a large quantum jump ushering in the yellow revolution increasing production of oilseeds from 11 to 22 million tonnes during the last one decade. The export of oil meal, oil-cake and minor oil has doubled. Likewise the dairy sector has witnessed a white revolution. In inland fisheries, the country has registered an unprecedented compound growth of 10 per cent during the nineties brining blue revolution. The growth in horticulture and poultry production has also been of a higher order (Anonymous, 1999). In spite of momentous increase in per capita availability of food grain in India, there is no surety of filling the number of empty stomachs multiplied in the years to come. The population of India has already crossed 104 billions and is still increasing alarmingly and that put a great pressure on the food grain production of India. On a rough reckoning it is an acceptable fact that our country has achieved a marvelous success in food grain production from a bare 51 million tonnes in 1951-52 to 212 million tonnes in the year 2003. In spite of this, India is still hungry. It is estimated that India would need to produce an additional 50 million tonnes of food grain by 2010 AD to feed its anticipated population in that year. As per latest information the projected milk production and per capita milk consumption were 80.00 MT and 210 grams per day for the year 2002 (Anonymous, 2001) and more than 100 million tonnes in 2009. Now we should plan for sustaining high milk production as well as increasing productivity of animals. Unless we concentrate our effort on these areas, the shortage of staple foods is likely to be one of the greatest problems. In other words, the need has been arisen for a "Second Green Revolution" in the country through the objective of providing as envisaged in the first "Green Revolution" had been achieved (Paroda, 2000). To achieve the best results from agriculture, Indian farmers will have to make the best use of available modern high production oriented agricultural technologies. This poses a major challenge not only for the policy makers but also more directly to the agricultural scientists, extension workers, agricultural educationist and farmers.

A definite approach to craft India as developed nation in the shortest possible time is realized by a rapid agricultural growth. To accomplish the admired dream of father of nation; Mahatma Gandhi, we have to work with commitment in taking over newer information and technology to our millions of farmers who work hard vigorously in the sun to feed the nation (Anonymous, 2000). The world is shifting very rapidly. Technology of today becomes outdated tomorrow and agriculture is not exception to this. Agriculture plays an imperative role in Indian economy. The recent advances in agriculture and animal husbandry have demonstrated potential for maximization of agricultural and milk productivity. Today the new developments and techniques have taken over the old ones. However, reality is also that expected numbers of the farmers still do not adopt such recently innovated technologies.

Our former prime minister, Shri Atal Bihari Vajpayee's new slogan *"Jai Vigyan"* in addition to *"Jai Javan"* and *"Jai Kishan"* has provided that the scientific community is not only a sense of pride but it has also big challenge to accomplish the task of doubling the food production during five years to come (Anonymous, 1999)*. Thus, the National Agricultural Technology Project (NATP) with an outlay of Rs. 1000 crores was dedicated recently to the Indian farmers with a firm belief that it would pave the way to doubling the farm production by the year 2010 (Anonymous, 2000)*.

The present age has been rightly termed as an "Information Age". People want adequate and authentic information as early as possible. Farmers as human beings are also anxious and become more desirous with the advancement in science and technology to know what is happening in the field of agriculture. Farmers have enthusiasm to obtain knowledge, particularly in the field of modern agriculture to become psychologically strong and favorable with necessary capacities to adopt modern methods of agriculture. To fulfill the expectations of the farmers, an efficient transfer of technologies is the sole solution.

An extensive extension and training programmes are highly essential to increase the productivity in agriculture. Moreover, many noteworthy changes in the field of agriculture have also been observed recently. These major changes include demand for agricultural product, growing population, economic liberalization, operationalization of GATT agreement, privatization of extension services, sustainability in farming practices, rapid growth in information technology, development of new production technology, role of NGOs and role of indigenous knowledge and experience of farmers. Most of these changes require competent farmers who are able to increase productivity and maintain sustainability of their farming system (Van den Ban and Hawkins, 1996). Therefore, the extension functionaries in coming years under changing scenario will have to play larger role as manager to discharge their responsibility for the fulfillment of the changing requirements of skills and knowledge of farmers (Tripathi, 2000).

Krishi Vigyan Kendra

Krishi Vigyan Kendra (**KVK**) is an admirable project of ICAR for testing and transmitting agricultural technology to link the space between production and productivity and to increase self-employment opportunities among farming communities. Krishi Vigyan Kendra is the grass root level vocational training institute deliberated for bridging the gap between the available technologies at the one end and their application for increased production at the other end. It provides a strong training support for bringing about production break through in agriculture. The main objectives of KVK are to identify the training needs; to conduct production oriented need based training, to organize different non-formal educational programmes and to develop practical training facilities. The specific objectives of KVK however may be many and varying depending upon the agro-ecosystem of the operational area, needs of the clients, resources available and the thrust required for increasing the economic status of the targeted group. Thus, each KVK is free to develop its specific objectives to gear up the immediate as well as long term needs of the operational area.

It is not an easy task to confirm the innovative character of Krishi Vigyan Kendra. Concerted and continuous efforts have to be made to realize the ideas of Krishi Vigyan Kendra, by the way of imparting tailor made skill training. Therefore, the successful planning and implementation of extension programmes require not only concerted efforts but also a combination of administrative and managerial skills besides a strong political will. The managerial component in the extension programme is quite complex and at the same time crucial too. It is necessary to create a team of 'turned on' managers who have the special sensitively, adaptability and motivation to undertake the task of Krishi Vigyan Kendra's activities. The management knowledge and practice may help in this task (Bhatia *et al.*, 1990).

The frontline extension system of the Indian Council of Agricultural Research (ICAR) presently includes a network of 504 farm science centers commonly known as Krishi Vigyan Kendras (KVKs) and 10 Trainers Training Centers (TTCs). The activities of the KVKs, which were initially confined to farmer's training and On Farm Demonstrations (OFDs), have been further extended to cover in-service training of extension personnel, so as to upgrade their skills in new technologies and the vocational training of farm youths.

The popular demand for establishment of more KVKs in the country is a testimony that the institutional mechanism of the KVK has served society in general and Indian agriculture in particular quite well. Targeted one KVK in each district of India and there will be a provision for more than one KVKs in the same district on demand basis. This is assured by our honorable prime minister Dr. Manmohan singh in the inaugural speech in National gathering of KVKs at ICAR held during 26th to 28th October - 2005 at ICAR, New Delhi (Anonymous, 2005).

Diversification to horticulture and value addition to agricultural commodities will hold the key for making Indian agriculture have competitive and profitable. Our thrust has to be providing support to disseminate technology in these areas to tap the potential to the fullest extent. Cheaper labour and extensive research and development support are our strengths. The increase in the production of food grains, oil seeds and dairy has firmed up our resolve to exploit the potential in developing horticulture sector (Anonymous, 1999).

For increasing agricultural productivity and ensuring household food and nutritional security, we need to provide new direction to the transfer of technology mechanism. This requires effective linkages among scientists working in different disciplines necessitate institutional restructuring including renewal/ reorientation of the existing system to meet the newer challenges.

To meet changing needs, it is essential to create a cadre of "Technology Agent" from the unemployed youths, who are better trained, equipped and committed to serve our farming community, while generating self employment for them. Some KVKs of India have already trained unemployed youths in vocational activities such as sericulture, mushroom production, poultry production, vermicomposting, fruits and vegetable processing, sheep and goat rearing, bio-technology, net and green house management, tissue culture, nursery management etc. It is believed that integration of science with rural development activities would enhance the functional capability and required impact of extension activities. Hence, the integration of KVKs, both for human resource development and frontline demonstration of technologies at the district level would add value to the process of technology transfer on a system made (Anonymous, 1999).

Thus, Krishi Vigyan Kendra is an institutional project of the ICAR to demonstrate technology generation, technology assessment, technology refinements and dissemination in the field of agriculture and allied sectors. The KVKs have been charged now to take up the responsibilities of technology evaluation and impact assessment, demonstration of technology on the farmer's field and update of the knowledge and skill for the farmers.

Genesis of the KVK

The ICAR constituted a committee in 1973 headed by Dr. Mohan Sinha Mehta of Sevamandir, Udaipur (Rajasthan) for working out a details plan for implementing the Krishi Vigyan Kendra. Based on the recommendation of this committee, Tamil Nadu Agricultural University, Coimbatore, started the first Krishi Vigyan Kendra in 1974 at Pondicherry. The success of KVK at many locations created a great demand for establishment of more KVKs in the each district of country. Nowadays, 630 KVKs are established and functioning in the country.

Objectives of the KVK

1. To plan and conduct survey of the operational area to inventories and characterize physical and human resources with special reference to identify technological graphs and training needs of the farming community.

2. To plan and conduct production oriented need based short and long duration training courses on the campus as well as off-campus with the priority on the weaker and poorer section.
3. To organize farm science clubs in rural areas for young farmers.
4. To develop and maintain the campus farm demonstration unit on scientific lines.
5. To compel all relevant recommendations for the district to the meaningfully utilized in the training courses and the follow up extension programmes.

Services of KVK ICT Network

Understanding the importance of information communication technology in dissemination of agricultural information, ICAR has also tried to link KVKs through Internet (Fig. 2 and 3). Total 250 KVKs have been decorated as e-linked KVKs in first phase allover the country. They are working for the betterment of the country in different ways, such as:

1. Package of practices and recommendation related to locally relevant governmental schemes etc.
2. Online services on localized weather forecasting, market information, latest recommendations on crop, animal, poultry, fishery etc.
3. Interactive voice response system on local weather forecast and market information.
4. Data base on Frequently Asked Questions (FAQs) by the farmers.
5. Sources of availability of inputs such as seed, seedlings, grafts, bio-fertilizers, bio-pesticides and agricultural implements with prices and user instructions.
6. Soil and water testing facilities and on-line reporting.
7. On-line consultancy to farmers through e-mail, voice mail and video-mail.
8. Event calendar of agriculturally important events in the region, training programmes, study tours, exhibitions and other activities.

9. Publication of newsletters and bulletins on KVK website.
10. Soil, water and plant diagnosis services.
11. Other services related to post harvest technology (PHT) and WTO.
12. Self help group and watershed development programme on-line services.
13. On-line marketing.

Management of Krishi Vigyan Kendra

The extension educationists expressed concern over the alarming knowledge practice gap in agriculture in the country. Experts estimated that about 70 per cent of the farm technologies available in India had not yet been adopted by the farmers as well as wide gap existed between potentials of researchers demonstrated on experimental farms and actual outcome at farmer's level (Anonymous, 1998). Thus, an efficient management is crucial for extension organization including Krishi Vigyan Kendra.

For effective management of irrespective nature of the organization, there are some fundamental principles and methods, which guide for effective functioning. The rise of management science and the increasing faith in its ability to solve myriad problems is an outstanding development of modern time.

As viewed by Koontz *et al.* (1980), management is the task of managers to secure and utilize inputs to the enterprise, transforming them through the managerial functions as planning, organizing, staffing, leading and controlling with the due considerations. The situation is altogether different in case of management of Krishi Vigyan Kendra as a non-business enterprise, which is exposed, to serious constraints and handicaps, not encountered in business enterprises.

Krishi Vigyan Kendra is an organization, which is concerned with social system within which, a number of groups of people are working/involving *i.e.* specialists, administrators, technicians, farmers, extension workers, labourers, other beneficiaries, etc. Each group has its own characteristics *viz.*, attitudes, traditions, values and form of behaviour. The people for whose benefits, the extension activities are designed may not be readily interested due to multiplicity of factors *viz.*, likeness, inhibitions, taboos, suspiciousness, reservations and prejudices (Bhatia *et al.*, 1990)*. It is therefore expected that manager of the KVK should have enough understanding of such social aspects to get best results out of KVKs.

The Revised Mandates of the KVK's

1. Conducting "**On Farm Testing**" for identifying technologies in terms of location specific sustainable land used system.
2. Organize the training to update the extension personnel with emerging advances in agricultural research on regular basis.
3. Organize short and long term vocational training courses in agricultural and allied vocations for the farmers and rural youth's with emphasis on "**Learning by Doing**" for higher production on farm and self-employment.
4. Organize **Front Line Demonstration** on various crops to generate production data and feedback information.
5. To act as a **Knowledge and Resource Centre** for farming Community at District Level.
6. Act as an **"Information Hub"** for farming community at District level.

Mandatory Activities of Krishi Vigyan Kendra

Based on mandate the following activities are performed by KVKs

1. On Farm Testing **(OFTs)**
2. In service training of extension functionaries
3. Vocational training of practicing farmers, farmwomen and rural youths
4. Front line demonstration **(FLDs)** and other extension activities.

The various said activities are performed every year by the Krishi Vigyan Kendra through six disciplines *viz.*, Extension Education, Agronomy, Horticulture, Plant protection, Animal Science, Agriculture Engineering etc. The specialists can change activities as per location specific need, which are most relevant taking into consideration natural resources and infrastructure facilities of the district. Moreover, it is a growing realization that an employee today is indeed a new breed. He appears to have different values, needs and motivation than his predecessor. He is a product of knowledge explosion. Thus, managing today's employee is difficult job. Further, the interpersonal relations with group and relationship between the groups also exist in organizations. The groups are mutually interdependent as they perform basically different functions and require different administrative approaches and management practices. A manager who understands the discipline of management will be more effective than a first rate manager with no more than minimum competence in managerial skills and tools.

The programme coordinator acts as an extension manager of Krishi Vigyan Kendra who is a key person, on whose shoulders, the effective functioning of Krishi Vigyan Kendra depends. The extension manager will have to perform an important task in guiding the change process. Some principles of management science should be considered which may increase the ability and the competence of extension managers as guides to the change process. Thus, many extension management functions have to be done properly by the programme coordinator of Krishi Vigyan Kendra as the head to get more productivity from the available resources. This book is addressed to those who love agricultural research, teaching and extension. Who love the sustainable agricultural development through effective and efficient Transferring of Agricultural Technologies (TOTs).This book will pave the means of those who are occupied with agriculture and rural developmental projects with enhancement in managerial ability, capability and potentialities. I hope this book will be of great use for extension workers.

Nikulsinh M. Chauhan

Contents

List of Figures

List of Plates

List of Tables

Chapter 1

Introduction

India's population has largely crossed 1 billion and by 1035 experts project that India will overtake China as the world's most populous country. At the same time India already holds first place in another category; the county is home to the world's largest number of poor people. Today estimated 350 million out of 1 billion Indians fall between the poverty line. This number represents a grim reality; while India claims self-sufficiency in production of grains, two-third of the Indian children under the age of 5 are malnourished. Infant mortality is 65 per 1000 live births, compared with 33 for China. Adult literacy rates of 65 per cent for men and 38 per cent for women fall far short of those in China 90 per cent for men and above 73 per cent for women.

Can India achieve its own economic miracle, pushing back poverty and stimulating a vibrant economy with opportunities for its entire people ? The task for India's policymakers is to recharge agricultural growth, enabling poor people to contribute their skills and energies to the country's economic activities to get the enhanced benefits of economic success.

Indian agricultural sector is vast and has continued to be the backbone of our economy. It supports 70 per cent of the country's population, which contributes 30-35 per cent of GDP and generates about 20 per cent of export earnings (Mehta, 1998). It provides employment for approximately 62 per cent of the workforce (Singh, 2000). It is indeed heartening that during the 55th year of India's independence, we could reap a record harvest of 208 million tonnes of food grains (Anonymous, 2002).

During the past four decades, spectacular progress has been achieved in agricultural production in the country, which has been well chronicled and universally recognized. In the oil seed sector,

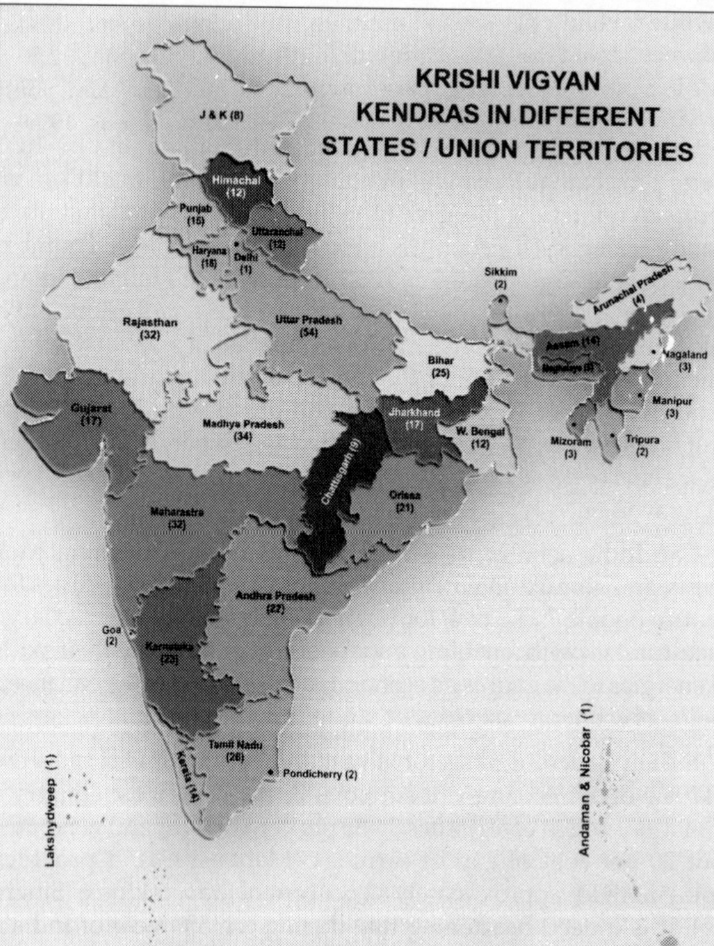

Plate 1 : Statewise scenario of KVKs in India

there has been a large quantum jump ushering in the yellow revolution increasing production of oilseeds from 11 to 22 million tonnes during the last one decade. The export of oil meal, oil-cake and minor oil has doubled. Likewise the dairy sector has witnessed a white revolution. In inland fisheries, the country has registered an unprecedented compound growth of 10 per cent during the nineties brining blue revolution. The growth in horticulture and poultry production has also been of a higher order (Anonymous, 1999).

In spite of momentous increase in per capita availability of food grain in India, there is no surety of filling the number of empty stomachs multiplied in the years to come. The population of India has already crossed 104 billions and is still increasing alarmingly and that put a great pressure on the food grain production of India. On a rough reckoning it is an acceptable fact that our country has achieved a marvelous success in food grain production from a bare 51 million tonnes in 1951-52 to 212 million tonnes in the year 2003. In spite of this, India is still hungry. It is estimated that India would need to produce an additional 50 million tonnes of food grain by 2010 AD to feed its anticipated population in that year. As per latest information the projected milk production and per capita milk consumption were 80.00 MT and 210 grams per day for the year 2002 (Anonymous, 2001) and more than 100 million tonnes in 2009. Now we should plan for sustaining high milk production as well as increasing productivity of animals. Unless we concentrate our effort on these areas, the shortage of staple foods is likely to be one of the greatest problems. In other words, the need has been arisen for a "Second Green Revolution" in the country through the objective of providing as envisaged in the first "Green Revolution" had been achieved (Paroda, 2000). To achieve the best results from agriculture, Indian farmers will have to make the best use of available modern high production oriented agricultural technologies. This poses a major challenge not only for the policy makers but also more directly to the agricultural scientists, extension workers, agricultural educationist and farmers.

A definite approach to craft India as developed nation in the shortest possible time is realized by a rapid agricultural growth. To accomplish the admired dream of father of nation; Mahatma Gandhi, we have to work with commitment in taking over newer information and technology to our millions of farmers who work hard vigorously

Plate 2 : Designers of Indian Agriculture

Plate 3 : Director General (ICAR) attending the National Conference

in the sun to feed the nation (Anonymous, 2000). The world is shifting very rapidly. Technology of today becomes outdated tomorrow and agriculture is not exception to this. Agriculture plays an imperative role in Indian economy.

The recent advances in agriculture and animal husbandry have demonstrated potential for maximization of agricultural and milk productivity. Today the new developments and techniques have taken over the old ones. However, reality is also that expected numbers of the farmers still do not adopt such recently innovated technologies.

Our former prime minister, Shri Atal Bihari Vajpayee's new slogan *"Jai Vigyan"* in addition to *"Jai Javan"* and *"Jai Kishan"* has provided that the scientific community is not only a sense of pride but it has also big challenge to accomplish the task of doubling the food production during five years to come (Anonymous, 1999)*. Thus, the National Agricultural Technology Project (NATP) with an outlay of Rs. 1000 crores was dedicated recently to the Indian farmers with a firm belief that it would pave the way to doubling the farm production by the year 2010 (Anonymous, 2000)*.

The present age has been rightly termed as an "Information Age". People want adequate and authentic information as early as possible. Farmers as human beings are also anxious and become more desirous with the advancement in science and technology to know what is happening in the field of agriculture. Farmers have enthusiasm to obtain knowledge, particularly in the field of modern

Plate 4 : Chairman, ICAR addressing KVK heads

Plate 5 : Secretary, DARE and DG, ICAR

agriculture to become psychologically strong and favorable with necessary capacities to adopt modern methods of agriculture. To fulfill the expectations of the farmers, an efficient transfer of technologies is the sole solution.

An extensive extension and training programmes are highly essential to increase the productivity in agriculture. Moreover, many noteworthy changes in the field of agriculture have also been observed recently. These major changes include demand for agricultural product, growing population, economic liberalization, operationalization of GATT agreement, privatization of extension services, sustainability in farming practices, rapid growth in information technology, development of new production technology, role of NGOs and role of indigenous knowledge and experience of farmers. Most of these changes require competent farmers who are able to increase productivity and maintain sustainability of their farming system (Van den Ban and Hawkins, 1996). Therefore, the extension functionaries in coming years under changing scenario will have to play larger role as manager to discharge their responsibility for the fulfillment of the changing requirements of skills and knowledge of farmers (Tripathi, 2000).

1.1 Agricultural Extension System

The productivity level in agriculture mainly depends to a large extent on what type of response a farmer gives to the various plans and strategies of agricultural extension evolved from time to time. The idea behind introducing the various extension programmes is to evolve to good response from the farmers (Singhi *et al.*, 1982).

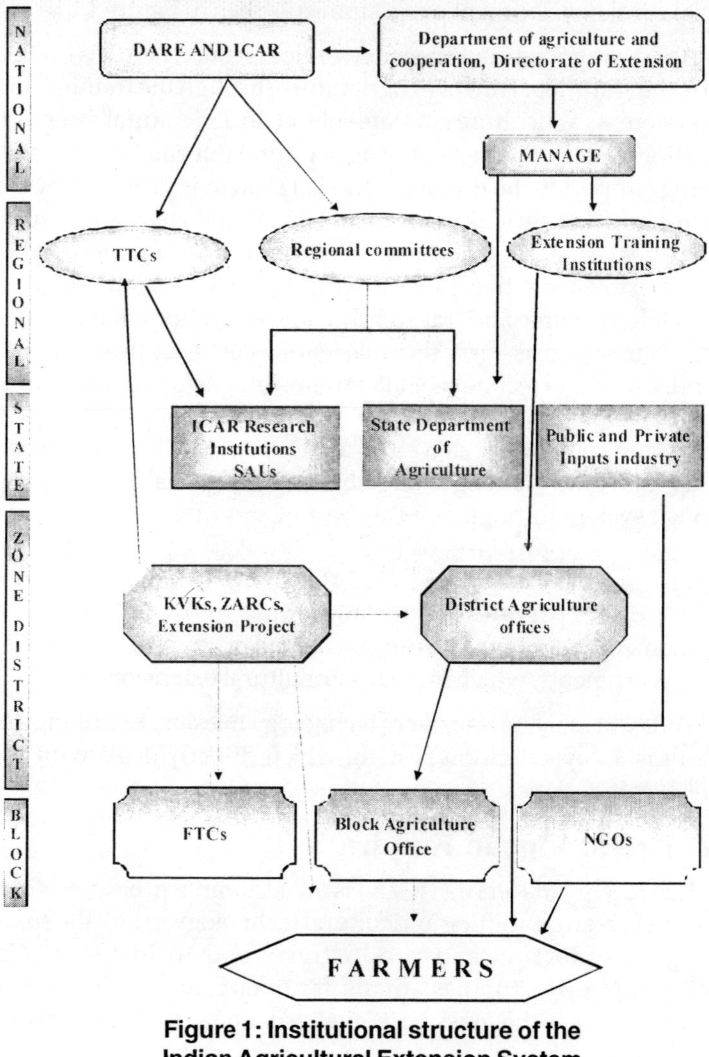

Figure 1: Institutional structure of the Indian Agricultural Extension System

Agricultural extension is element of the technology transfer system, which is mainly concerned with conveying information and knowledge of important technology from Researchers to Farmers. Broadly, there are four major components of the Indian Agricultural Extension System (IAES) or Transfer of Technology System at present. These components are agricultural extension service, agricultural extension education system, agricultural extension programmes and special rural development programmes as seen in Figure 1.

As far as agricultural extension service is concerned, it is carried out by the state department of agriculture through the training and visit system, as agriculture is a state subject. In agricultural extension education system, the transfer of agricultural information and skill training support to the training and visit system is provided by the indian council of agricultural research (icar) and state agricultural universities (saus). The extension system of icar and sau's undertakes first line extension activities, by this system they develop extension methodology, refine and transfer first line technology and provide feedback to scientists. First, the transfer of technology programmes, second, transfer of technology via training by krishi vigyan kendras on the principles of "teaching by doing" and "learning by doing" to the farming community as well as to the grass root level extension workers and third, training to the subject matter specialists of training and visit system through monthly workshops by icar and saus are the important approaches of this system. Agricultural extension projects/programmes are also undertaken by the inputs industries in the public/private sectors and some ngos. Special rural development programmes are implemented by the central as well as state governments, which include agricultural extension too.

Today in first line transfer of technology mission, Krishi Vigyan Kendra is a key extension system, which directly deals with the farming communities.

1.2 Krishi Vigyan Kendra

Krishi Vigyan Kendra (**KVK**) is an admirable project of ICAR for testing and transmitting agricultural technology to link the space between production and productivity and to increase self-employment opportunities among farming communities. Krishi Vigyan Kendra is the grass root level vocational training institute deliberated for bridging the gap between the available technologies at the one end and their application for increased production at the other end. It provides a strong training support for bringing about production break through in agriculture. The main objectives of KVK are to identify the training needs; to conduct production oriented need based training, to organize different non-formal educational programmes and to develop practical training facilities. The specific objectives of KVK however may be many and varying depending upon the agro-ecosystem of the operational area, needs of the clients, resources available and the thrust required for increasing the economic status of the targeted group. Thus, each KVK is free to develop its specific objectives to gear up the immediate as well as long term needs of the operational area.

It is not an easy task to confirm the innovative character of Krishi Vigyan Kendra. Concerted and continuous efforts have to be made to realize the ideas of Krishi Vigyan Kendra, by the way of imparting tailor made skill training. Therefore, the successful planning and implementation of extension programmes require not

Plate 6 : Director addressing the participants at KVK, Nashik

Plate 7 : Vice chancellor visits the KVK in-service training at Nashik

only concerted efforts but also a combination of administrative and managerial skills besides a strong political will. The managerial component in the extension programme is quite complex and at the same time crucial too. It is necessary to create a team of 'turned on' managers who have the special sensitively, adaptability and motivation to undertake the task of Krishi Vigyan Kendra's activities. The management knowledge and practice may help in this task (Bhatia *et al.*, 1990).

The frontline extension system of the Indian Council of Agricultural Research (ICAR) presently includes a network of 504 farm science centers commonly known as Krishi Vigyan Kendras (KVKs) and 10 Trainers Training Centers (TTCs). The activities of the KVKs, which were initially confined to farmer's training and On Farm Demonstrations (OFDs), have been further extended to cover in-service training of extension personnel, so as to upgrade their skills in new technologies and the vocational training of farm youths.

The popular demand for establishment of more KVKs in the country is a testimony that the institutional mechanism of the KVK has served society in general and Indian agriculture in particular quite well. Targeted one KVK in each district of India and there will be a provision for more than one KVKs in the same district on demand basis. This is assured by our honorable prime minister Dr. Manmohansinh in the inaugural speech in National gathering of KVKs at ICAR held during 26th to 28th October - 2005 at ICAR, New Delhi (Anonymous, 2005).

Diversification to horticulture and value addition to agricultural commodities will hold the key for making Indian agriculture have competitive and profitable. Our thrust has to be providing support to disseminate technology in these areas to tap the potential to the fullest extent. Cheaper labour and extensive research and development support are our strengths. The increase in the production of food grains, oil seeds and dairy has firmed up our resolve to exploit the potential in developing horticulture sector (Anonymous, 1999).

For increasing agricultural productivity and ensuring household food and nutritional security, we need to provide new direction to the transfer of technology mechanism. This requires effective linkages among scientists working in different disciplines necessitate institutional restructuring including renewal/ reorientation of the existing system to meet the newer challenges.

To meet changing needs, it is essential to create a cadre of "Technology Agent" from the unemployed youths, who are better trained, equipped and committed to serve our farming community, while generating self employment for them. Some KVKs of India have already trained unemployed youths in vocational activities such as sericulture, mushroom production, poultry production, vermicomposting, fruits and vegetable processing, sheep and goat rearing, bio-technology, net and green house management, tissue culture, nursery management etc. It is believed that integration of science with rural development activities would enhance the functional capability and required impact of extension activities. Hence, the integration of KVKs, both for human resource development and frontline demonstration of technologies at the district level would add value to the process of technology transfer on a system made (Anonymous, 1999).

Thus, Krishi Vigyan Kendra is an institutional project of the ICAR to demonstrate technology generation, technology assessment, technology refinements and dissemination in the field of agriculture and allied sectors. The KVKs have been charged now to take up the

Plate 8 : View of different KVKs of India

Plate 9 : View of different KVKs of India

responsibilities of technology evaluation and impact assessment, demonstration of technology on the farmer's field and update of the knowledge and skill for the farmers.

1.2.1 Genesis of the KVK

The icar constituted a committee in 1973 headed by dr. Mohan sinha mehta of sevamandir, udaipur (rajasthan) for working out a details plan for implementing the krishi vigyan kendra. Based on the recommendation of this committee, tamil nadu agricultural university, coimbatore, started the first krishi vigyan kendra in 1974 at pondicherry. The success of kvk at many locations created a great demand for establishment of more kvks in the each district of country. Nowadays, 630 kvks are established and functioning in the country.

1.2.2 Objectives of the KVK

1. To plan and conduct survey of the operational area to inventorize and characterize physical and human resources with special reference to identify technological graphs and training needs of the farming community.
2. To plan and conduct production oriented need based short and long duration training courses on the campus as well as off-campus with the priority on the weaker and poorer section.
3. To organize farm science clubs in rural areas for young farmers.
4. To develop and maintain the campus farm demonstration unit on scientific lines.
5. To compel all relevant recommendations for the district to the meaningfully utilized in the training courses and the follow up extension programmes.

1.2.3 Achievement of KVKs

At present 630 krishi vigyan kendra (kvks) are functioning full-fledged in the country in the service of indian farming. During recent year, these kvks have given benefits to 6,10,000 farmers and farmwomen, 1,30,000 rural youth and 68,708 in-service personnel by organizing total 30,517 training programmes. To accelerate dissemination process, kisan melas, discussion, farm advisory services and diagnostic film shows were arranged. The kvks also

Plate 10 : Transfer of technology thorugh 'Krishi Rath' in Gujarat state

Plate 11 : Inauguration of KVK by Hon. Governer in Gujarat

took up on farm testing (oft) of 607 farm technologies. A total of 32,340 front line demonstrations (flds), covering 1, 19,260 ha of area, were conducted to demonstrate the potential on newly released production technologies on oilseeds, pulses and other crops, besides this, they also carried out 3,337 demonstrations on livestock production and management. At KVKs 7,490.73 tonnes seeds of cereals, oilseeds, pulses and vegetables, spices, medicinal and aromatic plants, ornamental and plantation crops and forest species and 958670 livestock strains were produced and made available to the farmers. Advisory services were also provided to 3,42,634 farmers and other users. Significant impact of KVK technological interventions is also visible in villages in terms of rainwater harvesting, empowerment of rural and tribal women through income generation activities, dairying as sustainable enterprise for women, reclamation of sodic land, off-season vegetable production, backyard poultry and revival of ginger cultivation in different states (Anonymous, 2005)*.

1.2.4 Services of KVK ICT Network

Understanding the importance of information communication technology in dissemination of agricultural information, ICAR has also tried to link KVKs through Internet (Figures 2 and 3). Total 250 KVKs have been decorated as e-linked KVKs in first phase allover the country. They are working for the betterment of the country in different ways, such as:

1. Package of practices and recommendation related to locally relevant governmental schemes etc.
2. Online services on localized weather forecasting, market information,latest recommendations on crop, animal, poultry, fishery etc.
3. Interactive voice response system on local weather forecast and market information.
4. Data base on Frequently Asked Questions (FAQs) by the farmers.
5. Sources of availability of inputs such as seed, seedlings, grafts, bio-fertilizers, bio-pesticides and agricultural implements with prices and user instructions.
6. Soil and water testing facilities and on-line reporting.
7. On-line consultancy to farmers through e-mail, voice mail and video-mail.
8. Event calendar of agriculturally important events in the region, training programmes, study tours, exhibitions and other activities.

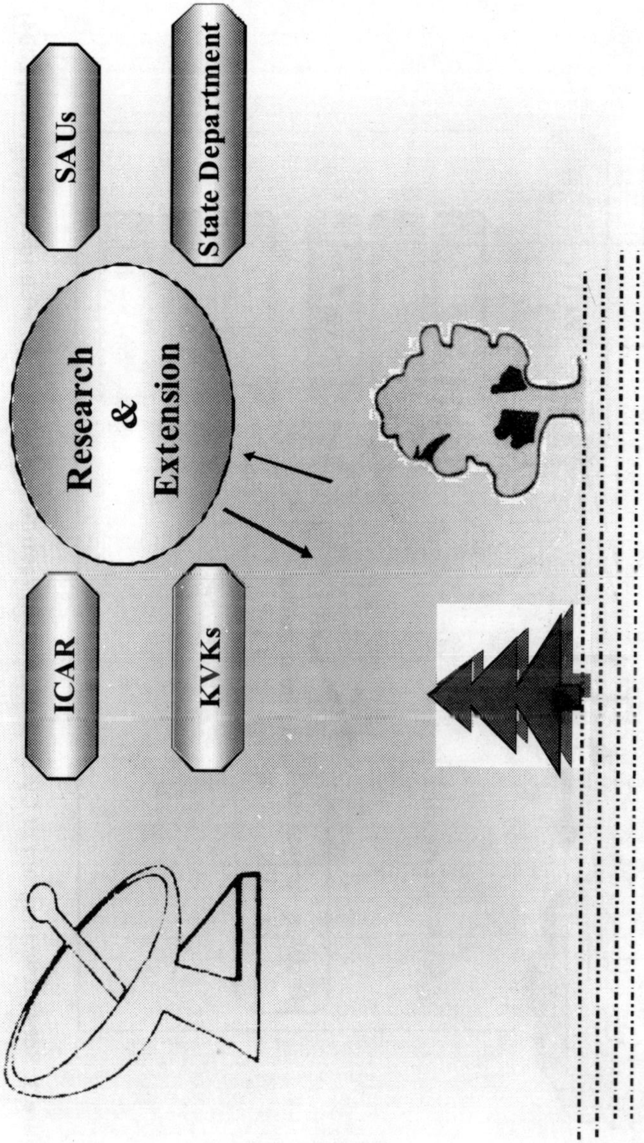

Figure 2: Rural Communication information network (Farm Net)

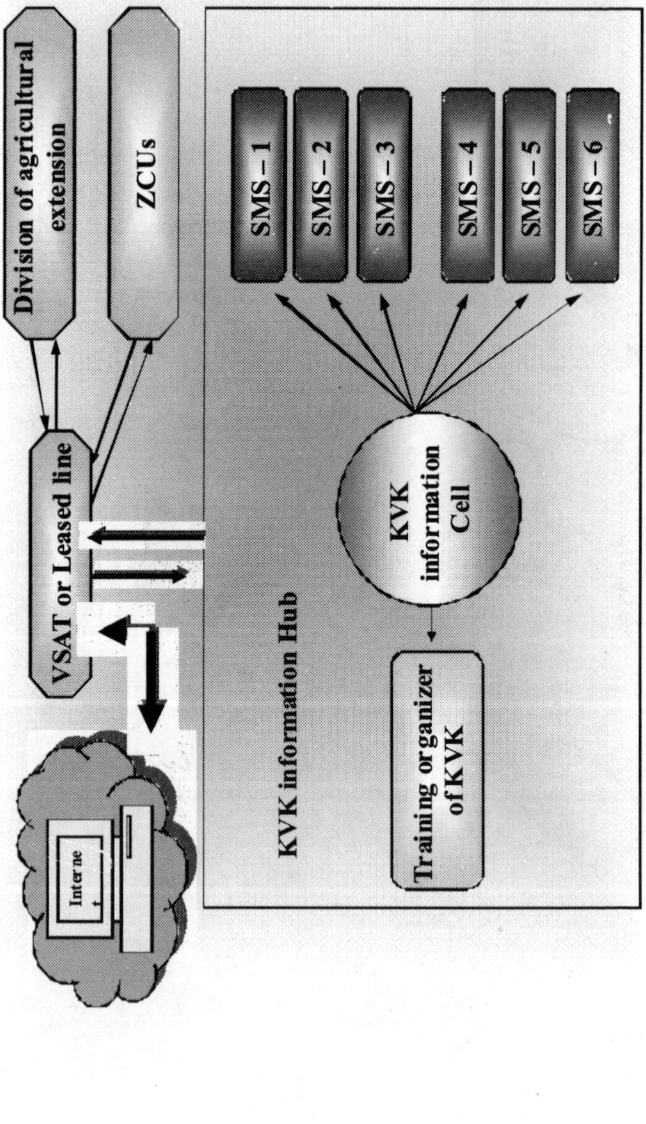

Figure 3: Schematic diagram of Cyber extension infrastructure to be build around KVK (Sawant, 2004)

9. Publication of newsletters and bulletins on KVK website.
10. Soil, water and plant diagnosis services.
11. Other services related to post harvest technology (PHT) and WTO.
12. Self help group and watershed development programme on-line services.
13. On-line marketing.

1.2.5 Management of Krishi Vigyan Kendras

The extension educationists expressed concern over the alarming knowledge practice gap in agriculture in the country. Experts estimated that about 70 per cent of the farm technologies available in India had not yet been adopted by the farmers as well as wide gap existed between potentials of researchers demonstrated on experimental farms and actual outcome at farmer's level (Anonymous, 1998). Thus, an efficient management is crucial for extension organization including Krishi Vigyan Kendra.

For effective management of irrespective nature of the organisation, there are some fundamental principles and methods, which guide for effective functioning. The rise of management science and the increasing faith in its ability to solve myriad problems is an outstanding development of modern time.

As viewed by Koontz *et al.* (1980), management is the task of managers to secure and utilize inputs to the enterprise, transforming them through the managerial functions as planning, organizing, staffing, leading and controlling with the due considerations. The situation is altogether different in case of management of Krishi Vigyan Kendra as a non-business enterprise, which is exposed, to serious constraints and handicaps, not encountered in business enterprises.

Krishi Vigyan Kendra is an organization, which is concerned with social system within which, a number of groups of people are working/involving *i.e.* specialists, administrators, technicians, farmers, extension workers, labourers, other beneficiaries, etc. Each group has its own characteristics *viz.*, attitudes, traditions, values and form of behaviour. The people for whose benefits, the extension activities are designed may not be readily interested due to multiplicity of factors *viz.*, likeness, inhibitions, taboos, suspiciousness, reservations and prejudices (Bhatia *et al.*, 1990)*. It is therefore expected that manager of the KVK should have enough understanding of such social aspects to get best results out of KVKs.

The Revised Mandates of the Kvk's

1. conducting "**On Farm Testing**" for identifying technologies in terms of location specific sustainable land used system.
2. Organize the training to update the extension personnel with emerging advances in agricultural research on regular basis.
3. Organize short and long term vocational training courses in agricultural and allied vocations for the farmers and rural youth's with emphasis on "**Learning by Doing**" for higher production on farm and self-employment.
4. Organize **Front Line Demonstration** on various crops to generate production data and feedback information.
5. To act as a **Knowledge and Resource Centre** for farming Community at District Level.
6. Act as an **"Information Hub"** for farming community at District level.

Mandatory Activities of Krishi Vigyan Kendra

Based on mandate the following activities are performed by KVKs

1. On Farm Testing **(OFTs)**
2. In service training of extension functionaries
3. Vocational training of practicing farmers, farmwomen and rural youths
4. Front line demonstration **(FLDs)** and other extension activities.

The various said activities are performed every year by the Krishi Vigyan Kendra through six disciplines *viz.*, Extension Education, Agronomy, Horticulture, Plant protection, Animal Science, Agriculture Engineering etc. The specialists can change activities as per location specific need, which are most relevant taking into consideration natural resources and infrastructure facilities of the district.

Plate 12 : Learning by doing at KVK

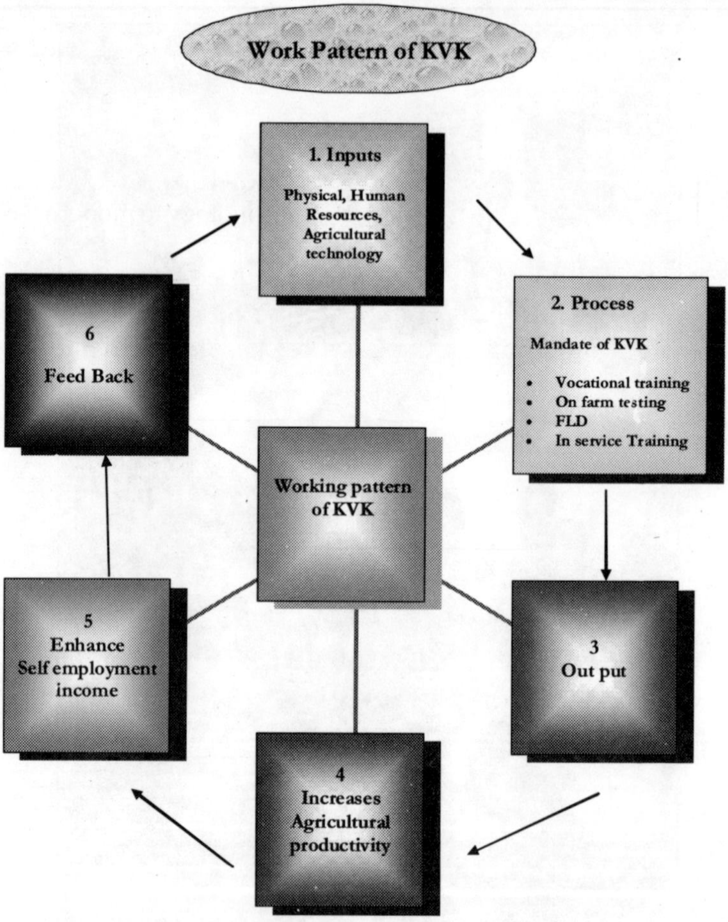

Figure 4: Working pattern of KVK

Moreover, it is a growing realization that an employee today is indeed a new breed. He appears to have different values, needs and motivation than his predecessor. He is a product of knowledge explosion. Thus, managing today's employee is difficult job. Further, the interpersonal relations with group and relationship between the groups also exist in organizations. The groups are mutually interdependent as they perform basically different functions and require different administrative approaches and management practices. A manager who understands the discipline of management will be more effective than a first rate manager with no more than minimum competence in managerial skills and tools.

The programme coordinator acts as an extension manager of Krishi Vigyan Kendra who is a key person, on whose shoulders, the effective functioning of Krishi Vigyan Kendra depends. The extension manager will have to perform an important task in guiding the change process. Some principles of management science should be considered which may increase the ability and the competence of extension managers as guides to the change process. Thus, many extension management functions have to be done properly by the programme coordinator of Krishi Vigyan Kendra as the head to get more productivity from the available resources.

1.3 Explanation of Problem

Krishi Vigyan Kendra is leading edged transfer of technology project of the ICAR. It is planned to create able human resource for agriculture development of the country. It helps in creating competent practicing farmers, farmwomen, rural youth and extension personnel. Understanding such importance of KVK, it is necessary to improve the operational quality of the KVK to get desired results. There is a need to analyze quantitative and qualitative functioning aspects of the KVK with a view to getting apprised strength and weakness. The KVK, which is started with set objectives, needs be assessed by evaluating managerial ability of its leader to understand its utility and effectiveness. It is, therefore, necessary to assess the qualities possessed by its heads or programme coordinators who are the key persons in effective functioning of centers to understand the works done by KVK in terms of agricultural development. Such study helps to locate the strong and weak points of managers of KVKs in overall functioning. It helps to give direction to the continued work besides brining confidence and satisfaction to the workers.

A creative and visionary leader with demonstrated integrity, enthusiasm, motivation, flexibility and confidence to act as programme coordinator has great chance to show the real impact of the KVK. The programme coordinator can establish and guide new research directions, lead technical and social research and development proposals, develop new technologies and capabilities and interact with in key clients. The programme coordinator as a key position holder should try to utilize opportunity to lead centre

Plate 13 : 'Seeing is believing', farmers visiting different demonstration units at KVK

Plate 14 : View of different vocational training programmes at KVK

and offer at the forefront of the emerging technology (Patel *et al.*, 2004).

The effectiveness of the Krishi Vigyan Kendras mainly depends on how it efficiently and successfully managed by its leader or the heads. Generally it is observed that the programme coordinators of the KVKs are postgraduates with specialization in different subjects of agriculture. They need to have sufficient experience in managing the multidisciplinary faculties of KVKs. The KVK programme coordinators have to perform multifold roles of planning, organizing, directing, communicating, controlling, supervising, coordinating, leading, coordination and decision-making.

These managerial qualities make them able to perform role as per the mandate of the KVK. By managing all those qualities they have to create conducive condition for effective training. The success of any KVK is directly or indirectly depends on the various managerial qualities of programme coordinators. Koontz and O'Donnell (1972) stated that the practice of managing like other practices will be better by making use of underlying organized knowledge and it is this knowledge whether cruder or advanced, whether exact or inexact that to the extent it is well organized clear and pertinent comprises a science. It was therefore, necessary to examine existing status of Programme coordinators of the KVKs. With a view to knowing this, a systematic study thought to undertake on "Managerial ability" of the programme coordinators.

1.4 Scope and Importance of Study

During The National Conference of KVKs of India, organized at ICAR, New Delhi from 27- 28 October, 2005, our honorable Union Minister of Agriculture, Government of India, Mr. Sharad Pawar expressed the salient features and the importance of KVKs in Indian agriculture and country's economy as a whole. He pointed out that after establishing the first KVK during 1974 at Pondicherry, we could establish another 276 KVKs till the end of the IX plan. He assured that GOI is actively considering to sanction atleast one KVK in each district of our country. He believed and expressed that we need motivated workers and vibrant KVKs to double India's food production, oilseeds production to reduce expensive imports of edible oils, harness potential of rainfed agriculture, transformed agricultural marketing and for the grand success of the second phase of the Green Revolution.

Plate 15 : Celebration of different days at KVK

These reflect the prime importance of the KVKs in the service of Indian agriculture. The agriculture scenario of the country has changed rapidly by shift in focus from production for home consumption to market oriented production, from staple cereals to cash crops, own labour use to purchased inputs, traditional technology to modern technology, zero marketable surplus to more marketable surplus, labour use to capital use, traditional knowledge based farming to knowledge and skill intensive farming, less risky, low-yielding and less remunerative to high yielding and income and profit yielding farming. In essence, there is a transition going on from family farming to surplus farming for markets and finally commercial farming. These all changes will be only feasible and adoptable through effective and efficient functioning of the KVKs. The National Horticulture Mission will be implemented by KVKs, (Anonymous, 2005)* whether it is public sector or managed by NGOs will be assisted the mission.

Thus, considering major importance, effectiveness of KVKs can be improved by providing valuable functional pattern with sound management upgradation only. To fulfill these dreams, this study will be quiet useful to the concerned persons. It is clearly understood that the managerial effectiveness will lead to organizational effectiveness and further organizational effectiveness will definitely support the national development. KVKs popularly known as "Krishi Mandirs" may be proved as firestone of Indian second "Green Revolution" programme. Accordingly the present study was conducted on the KVKs, which have successfully completed 5 years of functioning. The study was dealt with the management ability of the programme coordinators of the KVKs in India. The inferences drawn from this study would be helpful to the ICAR administrators, policy makers as well as zonal coordinating units in taking right decision and a way for improving the existing deficiencies. The study will be helpful in making necessary changes for increasing the effectiveness of the KVKs in the country. Many factors affect the extension management ability of the programme coordinator of KVKs, these factors may be of different natures, some may be organizational and social creating an environment where they have to work, while on the other hand, some factors are entirely personal due to their individual psychological make up. The results of the study will be very useful to ascertain the factors associated with management ability of programme coordinators of KVKs. The

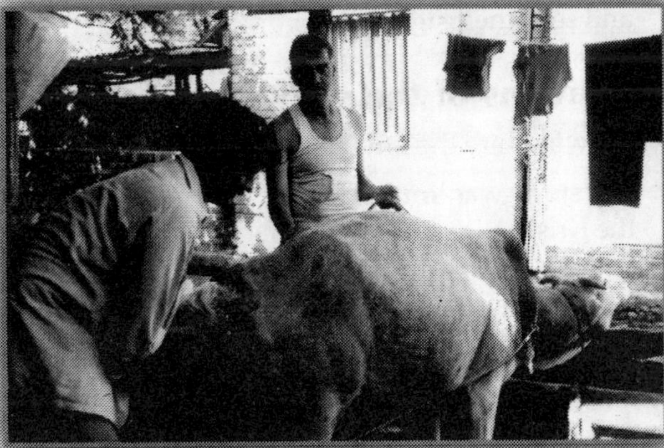

Plate 16 : Motivating efforts of KVK towards white revolution

information on key role played by some of the independent variables on extension management ability would also be of great use. The constraints faced by programme coordinators in performing extension management and suggestions offered by them to overcome some of the practical problems will also be useful in structuring effective strategy for future.

1.5 Objectives of Research

The present study was conducted with general objectives of measuring the extension management ability of programme coordinators as heads of Krishi Vigyan Kendras. The specific objectives of the study were as under:

1. To study personal, job related, organizational and socio-psychological characteristics of the Programme Coordinators of KVKs.
2. To measure the managerial ability of Programme Coordinators of KVKs.
3. To find out the relationship between selected independent variables and managerial ability of the Programme Coordinators of KVKs.
4. To estimate the functional relationship between managerial ability andvarious independent variables.
5. To study the constraints faced by Programme Coordinators of KVKs in the management of different activities of KVKs and seek their suggestions.

1.6 Limitations of the study

The followings are the limitations of the present study.

1. The study was limited to the Programme Coordinators of the Krishi Vigyan Kendra only.
2. Only some of the Programme Coordinators were studied.
3. The study was limited to measure only extension management ability of the Programme Coordinators of Krishi Vigyan Kendras.
4. The results in this study were based on the expressed opinions as well as ability to express own personality of

Programme Coordinators of selected Krishi Vigyan Kendras only.

5. The constraints faced by Programme Coordinators may be varied with host institution *viz.*, Government Organizations, NGOs, ICAR's etc.
6. Interpretations and conclusions were not generalized at macro level.

Chapter 2

Review of Literature

The main purpose of this chapter is to organize and present findings of the past studies related to the present investigation. It was revealed that very little research work has been carried out about the objectives under study. However, an effort was made to present a brief account of related literature, keeping in view the objectives of the study.

The findings by research workers, experts etc. at different lines in past are presented under following heads.

2.1 Management effectiveness/management ability of the respondents

2.2 Profile of the respondents

2.3 Relationship between independent variables and management ability

2.4 Constraints in management and suggestions for effective management

2.5 Recent reviewed literature

2.1 Management Effectiveness/Management Ability of the Respondents

Shah (2006) reported that majority of the internet user teachers were perceived by their students with high level of knowledge about subject matter (72.50 per cent), high level of motivation skill to involve students in teaching and learning process (59.17 per cent), high level of interest in teaching (70.83 per cent), highly punctual (79.17 per cent), highly confident in teaching (69.17 per cent), partially biased nature (90.83 per cent), highly skillful in using audio-visual aids (71.67 per cent) and with medium command on language (63.34 per cent), where as slightly more than half of them were perceived by their students with high level overall teaching ability (55.00 per cent), high degree of ability to control students (52.50 per cent) and with effective speech (50.00 per cent).

Gardebroek and Lansink (2003) studied the estimation of farm productivity differentials using panel data. Results revealed that specialized fig breeding farms that are characterized by high total factor productivity have more buildings and machinery than farms with low total factors productivity *i.e.* positive relation between farm productivity and managerial ability.

Patel and Patel (2003) reported that majority (70.77 per cent) of the heads of Krishi Vigyan Kendras had medium, followed by 17.69 per cent with low and 11.54 per cent with high level of management ability.

Trip *et al.* (2003) studied the managerial efficiency of commercial greenhouse growers. They said that nowadays the managerial decision making process has been given new attention, both in theoretical studies as well as in empirical research explaining differences in farm results. However, while critical to explaining efficiency managerial ability has been difficult to measure and therefore often ignored. This study attempted to measure managerial ability. It divided the decision making process into four steps: goal formulation, planning, monitoring and evaluation. The results showed positive associations between firm efficiency and the quality of decision making, indicating that this procedure has been potentially successful and it was more towards successful and a move towards successfully measuring a critical input.

Wu-Sx *et al.* (2003) studied the estimation and decomposition of technical efficiency for sugar beat farms and the results indicated that the average efficiency was 0.88 with 45 per cent of the farms in the sample exhibiting full efficiency and efficiency was positively correlated with managerial ability.

Alvarez and Arias (2002) stated that managerial ability has important implications for farm growth and results revealed that increasing farm size while holding managerial ability constant can be an important source of diseconomies of size.

Nathall (2001) reported that the psychology of decision making from a farm management perspective outlines, what psychology others for changing a person's attributes and considers the structure of a research programme aimed at developing methods for improving an individual's managerial ability.

Tolmay and Hernandez (2001) studied the resistance to biotic and abiotic stress in Triticeae and reported that response to the stress situation occurs independently of the managerial ability, skill and resource level of the producer.

Tripathi (2000) observed that the major area for development of management skills for extension personnel is management of man. For the management of man, the skill requirement on one hand is teamwork and on other hand, leadership, coordination, motivation, delegation, clarity of objectives, analysis of resources environment and developing strategic plans, project planning, timely action, review and feedback for individual effectiveness to manage teamwork.

Sah (1999) reported that the extension performance may be disappointing today but it may show resilience with introduction and institutionalization of new and innovative approaches in decision making, financing, provision of extension, building and maintaining linkage with other system and methods to deliver services through hybrid extension model.

Bhandari (1998) stated that the Indian saying *"Yatha Raja, Tatha Praja"* that is, if a leader is good and effective, every thing will go smoothly, is no exaggerations. But after all that is not easy to bring professionalism of management. At the same time organizational development in true sense is not possible without professionalism of management.

Anonymous (1998)* opined that everybody wants to be the "Best" and "Have Best", but the sources in terms of man, money, machine and material are limited. Thus, arouse the need for management that would results in efficient and effective utilization of various resources to fulfill the objective.

Kareem and Jayaramaiah (1997) studied problems perceived by the state in the planning and implementation of integrated Rural Development Programme as perceived by BDOs and VEOs. He reported that most important problems encountered by them were misuse of assets by beneficiaries, lack of managerial ability among beneficiaries to take care of the assets and for block officials it was lack of adequate transportation to reach beneficiaries efficiently.

Monappa and Saiyaddain (1997) defined an effective manager as one who properly develops in terms of basic intellectual abilities and the predispositions necessary for carrying out the task smoothly.

Verma (1997) reported components relevant for management of extension system *viz.;* authority structure, delegation of authority, decentralization of authority, control system. Authority with humanity, responsibility and accountability which if applied properly can make authority effective in administration and management.

Bhople and Palaspagar (1996) reported the significant relationship between supervision characteristics and managerial ability as well as Managerial ability and six management functions *viz.*, labour force, live stock numbers and area managed, educational status, length of service and experience

Van den Ban and Hawkins (1996) identified the conditions which an extension services must fulfill before it can work well. These are; good communication information, adaptation to changing environment, motivation of staff, flexibility, identification of management problem, participatory leadership, staff development, specialists, generalists, women extension workers and privatization of extension service for accountability. They further indicated that more systematic planning and careful evaluation of objectives/methods/results were internal parts of effectiveness of many extension organizations. Information produced in evaluation was an important tool for extension organization management.

Shivram and Dalal (1994) studied the effect of managerial ability on farm financial success and reported the relative importance of each management variable to farm financial success.

Unger *et al.* (1995) reported that the adoption of herbicide based conservation tillage in the southern United States Great plains was influenced by managerial ability and economic consideration.

Patel (1993) studied the research management ability of research scientists working at Head of the Department/Station/Project/Scheme in Gujarat Agricultural University. He concluded that the majority (71.67 per cent) of the heads exhibited medium level of research management ability.

Islam and Alli (1991) identified the important sub-areas for training on management of rural development programme officers namely; project planning, implementation, resource utilization, communication with human relation, motivation, reporting, feed back, coordination, cooperation, problem identification, solution, leadership and decision making.

Popat (1991) studied that productivity of subject matter specialist working under T and V system in Gujarat state. He reported that the majority (70.13 per cent) of the SMSs had moderate level of productivity. The field activities got highest mean productivity, followed by training and contact with research.

Prasad (1991) recommended that effective extension strategy should include; use of modern technology tools, training need assessment, new research information, close linkage between all the social scientists, redesigning extension system, use of feed back in planning, serious attitudinal problem, training on learning by doing, group action of NGOs, timely supply of inputs, proper extension methodology and government support for modernizing farming.

Samanta (1991) stated that effective extension education requires adequate human and financial resources with technical and management skills and knowledge associated with new technology to be disseminated to farmers. He also stressed on good personnel management, which is essential for extension programme success.

Bhatia *et al.* (1990) stated that successful integration of agriculture, industries and service/supportive sector in the integrated rural development programme depends on the proper planning, organizing, controlling and directing the various activities. The help/cooperation of target groups and implementing agencies are also highly needed. The cooperation of above mentioned sectors and the various agencies is a very difficult and challenging task. Only the management knowledge and practice may help in this task.

Patel (1990) concluded that a good teacher in agricultural college should have a sound and up-to-date knowledge of the subject, possessing atleast postgraduate degree in the respective subject and should be methodical in teaching. As far as personal qualities are concerned, he should be honest, punctual in his works, open minded, enthusiastic and having love for his professional as well as the clients.

Chari and Nandapurkar (1989) studied that the managerial ability of the respondents was highly significant with planning, organizing, human relation, supervision, communication, co-ordination and control. The two most important variables were supervision and human relationship, while the two least important were control and organizing.

2.2 Profile of the Respondents

2.2.1 Personal Variables

2.2.1.1 Age

Parmar (2005) revealed that majority (61.67 per cent) of the agricultural scientists were found with middle age group, followed by 38.33 per cent with old age and none of them were with young age.

Sukhadia (2005) found that exactly three-fifth (60.00 per cent) agricultural authors were found with middle age group, followed by old aged (34.55 per cent) and young aged (5.45 per cent), respectively.

Singh *et al.* (2004) revealed that majority of the respondents were young age group (49.41 per cent), while almost the same percentage (45.88 per cent) belonged to middle age group (36 to 45 years). There were hardly 5.00 per cent having more than 45 years of age. It implies that most of the Horticultural Development Officers were either young or of middle age group.

Patel (2001) indicated that 49.23 per cent of the training organizers of KVKs were found to be in old age group, followed by 34.62 per cent with middle age group while only 16.15 per cent of the Training organizers were from young age group.

Wase (2001) observed the majority of chilli grower managers (52.50 per cent) were in the age group of 36 to 50 years *i.e.* middle age category.

Borkar and Chimurkar (1996) reported that most of the scientists were in age group of 36 and above.

Patel (1993) indicated that majority (61.67 per cent) of the HODs were in old age group, followed by 33.33 per cent were in middle age group and 5.00 per cent were in young age group.

Patel (1992) observed that majority (60.00 per cent) of the respondents were found to be in the middle age group, followed by young age group (22.00 per cent) and old age group (18.00 per cent).

Popat (1991) reported that majority (74.02 per cent) of the Research Scientist working as Head of Department/Station/Project/ Scheme in GAU were in old age group followed by 22.08 per cent and 3.20 per cent of the respondents were in middle age group and young age group, respectively.

Malik (1988) revealed that majority (56.55 per cent) of the teachers in Gujarat Agricultural University were found to be in the middle age group, followed by young age group (26.26) and old age group (13.79per cent). He also indicated that more than 80.00 per cent of the teachers belong to young and middle age group.

2.2.1.2 Education

Parmar (2005) found that more than half (57.50 per cent) of the agricultural scientists were educated up to doctoral degree, followed by 39.17 per cent up to post graduate degree, while 2.50 and 0.83 per cent of them were graduates and post-doctorate, respectively.

Sukhadia (2005) indicated that exactly half (50.00 per cent) of the agricultural authors were educated upto M.Sc. degree, followed by 41.82 per cent and 8.18 per cent with Ph.D. and B.Sc. degree, respectively.

Singh *et al.* (2004) reported that majority (80.00 per cent) of HODs were possessed B. Sc. (Agri). degree, while 20 per cent of them were with M.Sc. (Agri).

Patel (2001) found that majority (71.54 per cent) of the Training organizers possessed doctorate degree, whereas 22.31 per cent and 6.15 per cent Training organizers were found with master and bachelor degree, respectively.

Wase (2001) showed that relatively higher proportion of chilli growers were primary school educated (29.17per cent), followed by 25.83 per cent with high school and 23.33per cent with college education, while only 5.84 per cent of them were illiterate.

Borkar and Chimurkar (1996) reported that farm scientists (83.33 per cent) had master degree and only 16.67 per cent possessed the doctorate degrees in their respective subjects of specialization.

Patel (1993) revealed that more than half (55.00 per cent) of the HODs were Ph. D and M. Sc. (36.67 per cent), whereas only 8.33 per cent of them were B. Sc.

Patel (1992) observed that more than half (53.00 per cent) of the respondents had their qualification as doctorate and 47.00 per cent of the respondents were M. Sc. Level.

Popat (1991) revealed that majority (80.52 per cent) of the SMSs working under T and V system in Gujarat State were graduate, followed by 10.19 per cent with agriculture diploma and 9.09 per cent with post graduation.

Khare *et al.* (1987) reported that the role performance level of SMSs with postgraduate qualification was judged higher than simple graduate SMSs.

2.2.1.3 Health

Patel (1993) indicated that more than half (55.00 per cent) of the research scientists working as Head of Department/Station/Project/Scheme in GAU were having very good health, whereas 38.33 per cent of them were having good health, only 6.67 per cent were having average health and none of them were having poor or very poor health.

2.2.1.4 Size of Family

Patel (2001) found that majority (70.00 per cent) of the Training Organizer of KVK had medium size of family; while equal number of respondents had small (15.38 per cent) and big family size (14.62 per cent).

Patel (1993) indicated that 40.00 per cent of the HODs had medium size of family, followed by 36.67 per cent and 23.33 per cent had small and big size of family, respectively.

Popat (1991) found that majority (62.33 per cent) of the SMSs working under T and V system in Gujarat State had medium size of family, while (32.47 per cent) of them had small size family and 5.20 per cent had big size of family.

2.2.1.5 Type of Family

Patel (1990) stated that majority (64.17 per cent) of the agricultural teachers were belonged to joint family, while 35.83 per cent belonged to nuclear family.

Thakkar (1986) found that majority (76.67 per cent) of the farmers were belonged to joint family.

2.2.1.6 Native Place

Shah (2006) reported that majority of the teachers working in AAU had rural background.

Patel (1993) reported that majority (81.67 per cent) of the Departmental heads were belonging from Gujarat state while remaining (18.33 per cent) of the respondents were belonging from other states.

Popat (1991) observed that majority (77.92 per cent) of the SMSs working under T and V system in Gujarat State were belonged to rural area.

2.2.2 Job Related Variables

2.2.2.1 Experience as a Head of Organization

Patel (2001) observed that majority (62.31 per cent) of the training organizers were in middle level of experience as KVK head followed by low (19.23 per cent) and high (18.46 per cent) level of experience as KVK head.

Patel (1993) revealed that majority (68.33 per cent) of the Departmental head had medium level of service experience as a head, followed by high (16.67 per cent) and low (15.00 per cent), respectively.

Patel (1990) indicated that about one half (49.35 per cent) of the respondents had medium level of service experience, followed by low (27.27 per cent) and high (28.38 per cent), respectively.

Popat (1991) studied the effects of service experience for particular post on productivity. He concluded that experience as SMS has positive but non- significant relationship with SMS's level of productivity.

Veerbhadraiah (1983) revealed that job performance of the Deputy Director was high as compared to Assistant Director of Agriculture. When performance of 27 specific items was considered individually, the Deputy Directors of Agriculture were observed to perform better than Assistant Directors of Agriculture.

Jaggi (1978) analyzed data for managers with service of five years and less in one category and service over five years in the other category. The results indicated that managers with shorter service would be less authoritarian.

Kholn and Schooler (1973) studied the relationship between man's occupational experience and his psychological functioning and suggested that these results form a continuing interplay between job and man in which the effects of job on man are far from trivial.

2.2.2.2 Total Experience

Shah (2006) reported that Majority (69.00 per cent) of the AAU teachers had high to medium level of teaching experience.

Parmar (2005) stated that 65.00 per cent of the agricultural scientists had medium service experience, whereas 18.33 per cent had less and 16.67 per cent had wide service experience.

Sukhadia (2005) observed that exactly three-fifth (60.00 per cent) of the agricultural authors had medium length of service; whereas 21.82 per cent had low and 18.18 per cent had wide length of service, respectively.

Singh *et al.* (2004) reported that almost half (49.41 per cent) of the HDOs had reasonable years of experience *i.e.* 13 to 22 years of service, while 45.88 per cent of them had less than 12 years of total service and remaining 5.00 per cent of them had rich experience.

Patel (2001) reported that majority (60.77 per cent) of the training organizers had medium level of total experience; followed by 23.08 per cent and 16.15 per cent with low and high level of total experience, respectively.

Borkar and Chimurkar (1996) revealed that scientists were having quite a large service experience of more than 16 years, which was spread through all major activities like teaching, research and extension in most of the cases.

Patel (1993) revealed that majority (71.67 per cent) of the department head had medium level of experience, whereas 16.66 per cent and 11.67 per cent of them had low and high level of experience, respectively.

Patel (1992) observed that about three-fifth (61.00 per cent) of the training organizers had medium level of teaching experience, followed by high (21.00 per cent) and low (18.00 per cent), respectively.

Popat (1991) observed that majority (75.32 per cent) of the SMSs had medium level of service experience, while 20.78 per cent and 3.90 per cent of them had low and high level of service experience, respectively.

Chaudhary *et al.* (1989) found that the service experience and job performance of ADOs were not associated with each other.

Sakaria (1988) reported that length of service of SMSs had positive and non-significant relationship with role performance.

Malik (1988) found that majority (66.21 per cent) of the teachers in Gujarat Agricultural University had medium experience (1.75 to 17.19 years) in research, followed by low experience (17.44 years) and high (16.55 years). The results indicated that two third of the respondents had medium experience in research.

Rao and Mohan (1986) reported that majority (67.23 percent) of the APAU teachers had average (13.3 to 27.5 years) total experience 16.36 per cent each had high (more than 27.5 years) and low (less than 13.3 years) total experience, respectively.

2.2.2.3 Training Received

Parmar (2005) reported that majority (78.33 per cent) of agricultural scientists had attained national level training, whereas 84.16 per cent of agricultural scientists and 37.50 per cent of agricultural scientists had never attained international and state level training, respectively.

Singh *et al.* (2004) reported that only 10.59 (per cent) of Horticultural Development Officers did not attend any training in the field of horticulture/vegetable crops, while almost 90 per cent of them had reasonable number of training attended. As high as 83.53 per cent of them had attended 1 to 2 weeks of training and remaining had undergone more than 15 days duration of training.

Patel (2001) observed that majority (76.00 per cent) of training organizers were found in medium level of training received, while 16.20 per cent and 7.60 per cent of them were found in high and low level of training exposure, respectively.

Singh (2001) reported that 32.52 per cent of the women scientists had attended 1 to 5 training courses within the country, while 1.84 per cent had attended 5 to 10 courses. Regarding training in foreign countries, it was found that 16.56 per cent of the women scientists had participated in 1 to 5 training courses abroad.

Patel (1993) revealed that majority (63.33 per cent) of the HODs had received training; out of which 28.33 per cent had attended less training (below average) and 35.00 per cent had attended more training (above average). About 36.67 per cent of the respondents did not receive any type in-service training.

Patel (1992) reported that majority (64.00 per cent) of the respondents were trained, out of which 31.00 per cent had attended less training (below mean), whereas 33.00 per cent had received more training (above mean). About 36.00 per cent of the respondents did not received any type of training.

Popat (1991) indicated that 25.97 per cent of the SMSs did not receive any type of in-service training, while 74.03 per cent of them were trained. He also concluded that, out of trained respondents, 38.96 per cent attended less training (below average), whereas 35.07 per cent had attended more training (above average).

Ardeshana (1989) reported that majority (72.30 per cent) of the teachers had low level of exposure of training, while only 27.70 per cent of the teachers had high level of exposure of training.

2.2.3 Organizational Variables

2.2.3.1 Span of Control

Patel (2001) found that great majority (90.00 per cent) of training organizers had medium level of span of control; followed by 6.15 per cent and 3.85 per cent respondents had high and low level of span of control, respectively.

Patel (1993) reported that majority (86.67 per cent) of HODs had medium level of span of control and 13.33 per cent of them had high level of span of control.

2.2.3.2 Staff Facility

Shah (2006) stated that Majority of the internet user teachers of AAU had average to adequate level of facility of spacious room to sit and work (96.60 per cent), computer (87.50 per cent), over head projector (83.30 per cent), compact disc writer (79.20 per cent), printer (77.50 per cent), internet connectivity (68.30 per cent) and departmental library facilities (60.00 per cent).

Sukhadia (2005) reported that less than half (45.45 per cent) of the agricultural authors had average facilities in their departments, whereas more than one-third (34.55 per cent) had adequate facilities and 20.00 per cent had inadequate facilities in their departments.

Patel (2001) found that 61.00 per cent of the training organizers felt that existing infrastructural facility in their KVKs was medium, followed by low and high level of infrastructural facility perceived by 20.00 per cent and 19.23 per cent, training organizer, respectively.

Patel (1993) found that majority (58.33 per cent) of HODs had perceived medium level of staff facility, whereas 30.00 per cent and 11.67 per cent of them had perceived high and low level of staff facility, respectively.

Patel (1992) indicated that majority (65.00 per cent) of the respondents felt that the feasibilities available to them were reasonable, whereas 21.00 per cent of them felt that the enough facilities were not available properly and 14.00 per cent felt that facilities available in the department were ample.

Ardeshana (1989) reported that majority (73.84 per cent) of the teachers had average facilities available in their departments. Whereas 15.40 per cent had inadequate facilities and only 10.76 per cent per cent had adequate facilities available in their department. He also concluded that there was positive and non-significant association between attitude of the teachers towards their job and facilities available in the department.

Reddy and Jayaramaiah (1990) reported that the facilities and resources were significantly correlated with effectiveness of VEOs.

Soni (1990) reported that majority (56.10 per cent) of the agricultural authors had average facilities in their departments; whereas 23.17 per cent had adequate facilities and 20.73 per cent had inadequate facilities in their departments.

Dakhore and Bhilegaonkar (1988) found that the proper infrastructural facilities had positive influence on the level of job satisfaction of veterinary extension personnel.

Malik (1988) concluded that more than two third (68.97 per cent) of the teachers expressed that the facilities available in their departments were average, whereas 15.86 per cent and 15.17 per cent had possessed adequate and inadequate facilities in their department, respectively.

2.2.3.3 Organizational Climate

Patel (2001) observed that more than half (56.92 per cent) of training organizers had medium level perception of their organizational climate, while an equal number of respondents (21.54 per cent) had low and high level of perception of their organizational climate.

Patel (1993) reported that majority (71.67 per cent) of the HODs had medium level of perception for their organizational climate, while almost equal number of them had low (13.33 per cent) and high (15.00 per cent) level of perception for departmental climate.

2.2.3.4 Interpersonal Communication

Patel (2001) reported that majority (60.77 per cent) of the training organizers had medium level of interpersonal communication, whereas 23.85 per cent and 15.38 per cent of them had high and low level of interpersonal communication.

Patel (1993) found that majority (76.67 per cent) of the HODs had medium level of interpersonal communication, where 13.33 per cent and 10.00 per cent of them had low and high level of interpersonal communication, respectively.

Popat (1991) observed that majority (71.42 per cent) of the SMSs had medium level of interpersonal communication, while equal number (14.29 per cent) of them had high and low level of interpersonal communication, respectively.

Reddy and Sinha (1990) reported that in case of communication between professor and the assistant professor cadre, the communication process was possible through the associate professors but their role was not more than merely handling over the communication material. They are in fact quite seniors and they feel that they should have been an effective agent rather than merely a postal agent.

Samantha (1985) reported that scientists were not fully satisfied with the communication systems on research, personnel and administrative aspects prevailing in the ICAR institutes.

2.2.4 Socio-psychological Variables

2.2.4.1 Personality of the Head

Shah (2006) reported that Majority (74.20 per cent) of the internet user AAU teachers had medium level of perception about their professional development and having good relation among staff members.

Patel (2001) revealed that more than two-third (69.23 per cent) of the training organizers exhibited extrovert personality, whereas 30.77 per cent exhibited introvert type of personality.

Patel (1993) found that majority (86.67 per cent) of the HODs exhibited extrovert personality, whereas 13.33 per cent had introvert personality.

Bhimjiani (1978) concluded that an extension worker; the agricultural extension officers who were free from worries about dealing with the people and who had developed self confidence of working on group situations or as a member of the team were likely to be more successful.

Chaya (1974) concluded that effective teachers had better personality adjustment.

According to Stagner (1956) each person perceives the world in a manner slightly move different than others. These differences are results of unique individual personality and psychological make up. A job, which is satisfying and motivating to one person, may be boring to others.

2.2.4.2 Attitude towards Extension Work

Shah (2006) observed that more than half (58.34 per cent) of the internet user AAU teachers had impartial attitude, 35.00 per cent had favorable and 6.66 per cent of them had unfavorable attitude towards teaching activity.

Parmar (2005) reported that majority (75.83 per cent) of the teachers of GAU had neutral attitude towards publication, followed by unfavourable and favourable with 15.00 and 9.17 per cent, respectively.

Patel (2001) observed that majority (67.69 per cent) of the training organizer had favourable attitude towards extension work, followed by less favourable (16.92 per cent) and highly favourable (15.39 per cent) attitude towards extension work.

Patel (1993) observed that majority (70.00 per cent) of the HODs had medium intensity of favourable attitude towards research work, followed by low (16.67 per cent) and high (13.33 per cent) intensity of favourable attitude towards research work.

Patel (1992) found that majority (65.00 per cent) of the teachers had medium level of attitude towards teaching, followed by low (21.00 per cent) and high (14.00 per cent) level of attitude towards teaching, respectively.

Popat (1991) observed that majority (76.62 per cent) of the SMSs had medium level of attitude towards extension work, followed by high (15.58 per cent) and low (7.80 per cent) level of attitude towards extension work, respectively.

Rani (1985) reported that majority of the teachers were having favourable attitude towards teaching, research and extension. However she further revealed that only the attitude towards teaching was significantly associated with teaching productivity.

2.2.4.3 Perceived Work Load

Shah (2006) observed that slightly more than half (53.34 per cent) of the internet user AAU teachers had perceived heavy workload.

Sukhadia (2005) indicated that slightly more than half (51.82 per cent) of the agricultural authors had heavy workload, followed by average (46.36 per cent) and only 1.82 per cent had light workload.

Patel (2001) revealed that more than half of the training organizers (58.46 per cent) had perceived medium level of workload, followed by heavy (40.00 per cent) workload and only 1.54 per cent of them had perceived light workload.

Patel (1993) found that majority (71.67 per cent) of the HODs had perception of heavy workload, whereas 26.67 per cent had perception of medium workload and only 1.66 per cent of them had perception of light workload.

Patel (1992) reported that exactly three-fifth (60.00 per cent) of the teachers had average workload, followed by heavy (38.00 per cent) and only 2.00 per cent of them who had light workload.

Ardeshana (1989) indicated that the majority (53.74 percent) of the teachers had heavy workload, followed by 44.62 per cent with average and 1.54 per cent with light workload.

Soni (1990) indicated that majority (54.88 per cent) of the agricultural authors had heavy workload, followed by average (42.68 per cent) and only 2.44 per cent had light workload.

Malik (1988) found that majority (53.10 per cent) of Gujarat Agricultural University teachers had perceived average workload, followed by heavy (37.93 per cent) and too heavy (8.97 per cent). It was found that there was none of the teachers who had very light and light workload. This indicated that majority of the teachers had average and heavy workload.

Rao (1986) reported that majority (57.27 per cent) of the PAU teachers felt that their workload was up to the average degree, where as 31.81 per cent felt that it was heavy, 6.36 per cent said that it was too heavy, 3.36 per cent felt it light and a fraction of them (0.90 per cent) felt that their workload was very light.

2.2.4.4 Job Involvement

Shah (2006) stated that majority (85.00 per cent) of the AAU teachers had medium to high level of involvement in overall internet exposure.

Patel (2001) found that majority (62.31 per cent) of the training organizers were moderately involved in their jobs, followed by high level (20.77 per cent) and low level (16.62 per cent).

Patel (1993) observed that three-fourth (75.00 per cent) of the HODs exhibited medium level of involvement in their job, whereas 15.00 per cent of them had high level of job involvement and only 10.00 per cent had low level of job involvement

Popat (1991) reported that four fifth (79.22 per cent) of the SMSs exhibited medium level of involvement in their job, whereas 14.29 per cent of them had low level of job involvement and only 6.49 per cent had high level of job involvement.

Veerbhadraiah and Jalihal (1983) observed that some common factors like Job centrality, Duty conscious, striving perfection, indifferent response to work, disgusted with work and inconsistent were responsible for underlying job involvement of Deputy Directors and Assistant Directors of Agriculture.

2.2.4.5 Job Stress

Shah (2006) concluded that half (50.84 percent) of the internet user AAU teachers had medium level of job stress.

Patel (2001) revealed that about two-third of the training organizers (69.23 per cent) expressed medium level of job stress, whereas 17.69 per cent and 13.08 per cent respondents had experienced high and low level of job stress, respectively.

Patel (1993) found that majority (70.00 per cent) of the HODs experienced medium level of job stress, while equal number (15.00 per cent) of them had high and low level of job stress.

Popat (1991) indicated that majority (63.64 per cent) of the SMSs experienced medium level of job tress, followed by low (19.48 per cent) and high (16.88 per cent) level of job stress, respectively.

Igodan (1987) reported that majority (80.00 per cent) of the extension agents of Ohio State (U.S.A.) were found to experience a low to moderate level of burnout, while less than 20.00 per cent of all agents scored in the upper third of the scores indicating a high level of burnout.

Talukdar (1984) revealed that majority (54.90 per cent) of the ADOs experienced medium level of stress in their job lives. A significant number of them (26.00 per cent) also expressed a high level of job stress.

2.3 Relationship between Independent Variables and Management Ability

2.3.1 Personal Variables

2.3.1.1 Age and Management Ability

Shah (2006) stated that the age of AAU teachers was observed negatively significant the efficient involvement of teachers in managing internet.

Patel (2001) reported that age of the training organizer was not significantly associated with their extension management ability.

Shivram and Dalal (1994) observed significant but negative relationship between age and management of adoption process of gram production Technology in Haryana.

Patel (1993) revealed that there was positive and significant relationship between age of the research scientists and their research management ability.

Patel (1992) found that age of the respondents had positive and significant relationship with teaching effectiveness of the teachers.

Popat (1991) observed that age had non-significant relationship with productivity of the subject matter specialists.

Reddy and Jayaramaiah (1990) observed non-significant correlation between age and the productivity of VEOs.

Nikhade and Kitey (1984) found significant relationship between age of VLWS and their performance. Similar findings were also reported by Sharma *et al.* (1988).

2.3.1.2 Education and Management Ability

Shah (2006) further stated that academic qualification and teaching experience were found non-significant with the internet exposure of the AAU teachers.

Patel (2001) observed that academic qualification of the training organizers was positively and significantly related with their extension management ability.

Jondhale *et al.* (2000) reported that education of trained and untrained farmers established significant and positive relationship with the managing ability to adopt recommended groundnut cultivation.

Patel (1993) revealed that there was non-significant relationship between education and research management ability of the research scientists.

Popat (1991) reported that academic qualification was not significant with productivity of subject matter specialists.

Padheria (1989) concluded in a study on extension productivity that academic qualification was positively and significantly correlated with extension productivity.

Sharma *et al.* (1988) revealed that the role performance of Rural Agricultural Extension Officers of T and V system was significantly and positively correlated with their education.

Khare *et al.* (1987) showed that the academic qualification of SMSs was significantly associated with their role performance level.

Singh *et al.* (1982) reveled that education of trained farmers of Krishi Vigyan Kendra had positive and significant relationship with their knowledge and management of adoption level of recommended practices.

2.3.1.3 Health and Management Ability

Patel (1993) revealed that there was non-significant relationship between health of the Research scientists and research management ability.

2.3.1.4 Size of Family and Management Ability

Patel (2001) concluded that there was non-significant relationship between size of family of the training organizers and extension management ability

Patel (1993) observed that there was non-significant relationship between size of family and research management ability of the Research scientist.

Popat (1991) reported that size of family had non- significantly associated with the productivity of subject matter specialists.

2.3.1.5 Type of Family and Management Ability

Sudesh (2005) concluded that there was non-significant relationship between type of family and attitude towards fruit and vegetable training.

Shivram and Dalal (1994) studied the effect of managerial ability on farm financial success and reported that there was positive and significant relationship between managerial ability and type of family of farm growers.

England *et al.* (1971) reported that in any organization, the attitude of employees towards their work and type of family has a great bearing on the success or failure.

2.3.1.6 Native Place and Management Ability

Shah (2006) reported that native place of the AAU teachers found negative association with their overall internet exposure

Patel (1993) observed that there was non-significant relationship between belonging state of the Research scientist and their research management ability.

Popat (1991) reported that there was no relationship between rural-urban backgrounds of subject matter specialists and their level of productivity.

2.3.2 Job Related Variables

2.3.2.1 Experience as a Training Organizers in KVK and Management Ability

Patel (2001) concluded that there was no significant relationship between experience as a head or the training organizers and extension management ability.

Patel (1993) observed that there was non-significant relationship between experience as a research head of head of the organizations and their research management ability.

Patel (1992) observed that there was significant relationship between teaching effectiveness of teachers and their teaching experience.

Popat (1991) studied that there was no relationship between service experience as subject matter specialist of SMSs and their level of productivity.

Chaudhary *et al.* (1989) found that the service experience and job performance of ADOs were not associated with each other.

Rani (1985) reported that there was no significant association between experience and productivity of agricultural scientists.

2.3.2.2 Total Experience and Management Ability

Patel (2001) concluded that there was non-significant relationship between total experience as a head of KVK/the training organizers and extension management ability.

Patel (1993) observed that there was non-significant relationship between total service experience of heads of the organizations and their research management ability.

Popat (1991) studied that there was no relationship between service experience of subject matter specialists and their research management ability.

Padheria (1989) reported that total experience had positive and significant relationship with their extension productivity.

Sakaria (1988) reported that length of service of SMSs had positive and non-significant relationship with their role performance.

Khare *et al.* (1987) showed that the service experience of SMSs was non-significantly correlated with their role performance level.

2.3.2.3 Training Received and Management Ability

Patel (2001) concluded that there was non-significant relationship between training received by the training organizers and their extension management ability.

Patel (1993) observed that there was non-significant relationship between training received by research heads and their research management ability.

Patel (1992) indicated that there was no relationship between teaching effectiveness of teachers and their training received.

Popat (1991) studied that there was no relationship between training received by subject matter specialists and their research management ability.

Padheria (1989) reported that training received by the extension workers had non-significant relationship with their extension productivity.

Sakaria (1988) reported that total training received by the SMSs had non-significant relationship with their role performance.

Rani (1985) reported that training received by the APAU teachers had no relationship with extension productivity.

2.3.3 Organizational Variables

2.3.3.1 Span of Control and Management Ability

Patel (2001) indicated that span of control had non-significant relationship with extension management ability of the training organizers.

Patel (1993) reported that there was no relationship between span of control of the heads of the research organizations and their research management ability.

2.3.3.2 Staff Facilities and Management Ability

Patel (2001) found that there was non-significant relationship between staff facility and extension management ability of the training organizers.

Patel (1993) found that there was non-significant relationship between facility of the staff with head and their research management ability.

Patel (1992) indicated that there was no relationship between facilities available in department of teachers and their level of teaching effectiveness.

Reddy and Jayaramaiah (1990) reported that the facilities and resources were significantly correlated with effectiveness of VEOs.

Ardeshana (1990) concluded that there was positive and non-significant association between the facilities available in the department and the efficiency of the teachers.

Dakhore and Bhilegaonkar (1988) found that the proper infrastructural facilities had positive influence on the level of job satisfaction of veterinary extension personnel.

Rani (1985) observed that facility provided to the APAU teachers had non-significant relationship with the extension productivity.

2.3.3.3 Organizational Climate and Management Ability

Patel (2001) observed that there was non-significant relationship between organizational climate of the training organizers and their extension management ability.

Patel (1993) reported that there was non-significant relationship between organization climate of the heads of the departments and their research management ability.

Padheria (1989) reported that departmental climate experienced by the extension personnel had no relationship with the extension productivity.

Murthy and Somasundaram (1989) reported that organizational climate was significantly related with role performance of VEWs.

Rani (1985) observed that there was non-significant relationship between organizational climate experienced by the APAU teachers and extension productivity.

2.3.3.4 Interpersonal Communication and Management Ability

Patel (2001) reported that interpersonal communication had non-significant relationship with extension management ability of the training organizers.

Patel (1993) found that there was no relationship between interpersonal communication of heads of research organizations and their research management ability.

Padheria *et al.* (1992) reported that infrastructural facility and interpersonal communication were the most important variable affecting scientific productivity of academic staff.

Popat (1991) observed that there was non-significant relationship between interpersonal communication and level of productivity of research scientists.

Reddy and Jayaramaiah (1990) revealed that interpersonal communication was found significantly correlated with productivity of VEOs.

Laharia (1978) observed that the interpersonal communication had positive correlation with the productivity of agricultural scientists.

2.3.4 Socio-Psychological Variables

2.3.4.1 Personality of the Head and Management Ability

Patel (2001) revealed that there was negative significant relationship between personality of the training organizers and their extension management ability.

Patel (1993) found that there was non-significant relationship between personality of the heads of research organizations and their research management ability.

Bhimjiani (1978) stated that significant relationship between the personality factor and the performance of the agricultural extension officers.

2.3.4.2 Attitude Towards Extension Work and Management Ability

Shah (2006) reported that attitude towards teaching, professional development, job satisfaction, teaching aptitude of teacher and level of aspiration of the AAU teachers were observed non-significantly correlated with their level of internet exposure.

Patel (2001) observed that attitude towards extension work had positive and significant relationship with extension management ability of training organizers. Patel (1993) observed that there was no relationship between attitude of the HODs towards delegation of authority and their research management ability.

Patel (1992) found that attitude towards teaching had positive and significant relationship with teaching effectiveness of the agricultural teachers.

Popat (1991) reported that attitude towards extension work had no relationship with level of productivity of SMSs.

Malik and Patel (1990) concluded that attitude towards teaching was found significantly related with degree of satisfaction on the job of the teachers of GAU, Anand campus.

Reddy and Jayaramaiah (1990) showed that attitude of VEOs towards T and V system had significant correlation with their productivity.

Murthy and Somasundaram (1989) revealed that attitude of village extension officers towards T and V system was found non-significant with role performance.

Dakhore and Bhilegaonkar (1988) reported that attitude of veterinary extension personnel towards job had positive and significant relationship with degree of satisfaction on their job.

Rani (1985) revealed that attitude of the APAU teachers towards teaching was significantly associated with teachers productivity.

England *et al.* (1971) concluded that in any organization, the attitude of employees towards their work has a great bearing on the success or failure.

2.3.4.3 Perceived Work Load and Management Ability

Patel (2001) revealed that perceived workload by training organizers had non-significant relationship with extension management ability.

Patel (1993) found that workload perceived by the HODs had negative and non-significant relationship with research management ability of the heads of the organization.

Patel (1992) reported that there was no relationship between perception of workload of agricultural teachers and their level of teaching effectiveness.

Padheria (1989) reported that there was no relationship between workload perceived by the extension personnel and the extension productivity.

Rani (1985) observed that perceived workload by APAU teachers had non-significant relationship with their extension productivity.

2.3.4.4 Job Involvement and Management Ability

Patel (2001) found that there was non-significant relationship between job involvement of the training organizers and extension management ability.

Patel (1993) observed that job involvement of the HOD had non-significant relationship with their research management ability.

Popat (1991) reported that there was no relationship between job involvement of the SMSs and their level of productivity.

Murthy and Somasundaram (1989) concluded that job involvement of village Extension officers was found to be positive and significantly correlated with their role performance.

Reddy and Jayaramaiah (1988) reported that job involvement of VEOs was found to be not significant with job effectiveness.

Veerbhadraiah (1983) revealed that there was significant relationship between the job involvement of Deputy Directors and Assistant Directors of Agriculture and their job performance.

2.3.4.5 Job Stress and Management Ability

Patel (2001) revealed that there was non-significant relationship between job stress and extension management ability.

Patel (1993) found that there was no relationship between job stress of research managers and their research management ability.

Popat (1991) indicated that there was no relationship between job stress of subject matter specialists and their level of productivity.

Murthy and Somasundaram (1989) concluded that job stress was found non-significant with role performance of VEOs.

Talukdar (1984) stated that ADOs job stress exhibited no relationship with their productivity.

2.4 Constraints in Management and Suggestions for Effective Management

Patel (2001) reported that major constraints faced by the heads of the farmers training organization were (i) insufficient staff facility especially technical (ii) disciplines and qualification of the technical staff were not maintained (iii) inadequate facility in respect to demonstrations unit/training laboratory as well as infrastructure (iv) insufficient grant and delay in release of grant (v) inadequate power and authority with the training organizers (vi) lack of vehicle facility especially carrying extension work at village level (vii) modern information and communication technology were not fully available with KVKs which resist the modern development in management of extension work (viii) there was communication gap between research institution and extension organizations as well as non-compliance of technical recommendations. Being situated at remote areas, major farmers training centers faced the poor rural infrastructural and educational facilities, which indirectly influenced management of extension work. (ix) work motivation/ incentives with human relation angle was found absent in extension field.

They suggested several measures on above constraints *viz.* (i) vacant posts especially technical including training organizers should be filled up immediately (ii) discipline as well as qualification should be maintained while appointing technical staff. (iii) demonstration unit/training laboratory as well as infrastructural premises should be well equipped. (iv) sufficient grant especially for extension activities should be released timely. (v) power and authority should be delegated to the training organizers adequately, considering their responsibilities. (vi) the centers should be restricted to perform only extension tasks. (vii) modern information and communication technology such as computer, internet fax, zerox, mobile phone etc. should be made available. (viii) a strong coordination and linkage on permanent basis should be established between research institutions and centers. (ix) all the necessary infrastructural facility and educational facility for the staff should be made easily available. (x) There should be a provision for work motivation and incentives for the staff.

Reddy and Rao (1998) concluded that scientist commonly expressed constraints about scientific equipment, training and career development, availability of budget in time, library, laboratory, information system, transport and data analysis system in order of priority.

Pal and Singh (1997) observed that extension system is fraught with many problems and misconcepts *viz.* (i) department of agriculture feels encroachment of its domain by research system, while researches feel gap in transfer of technologies to farmers (ii) lack of operational funds, (iii) ritualistic nature of extension approach (iv) inadequate training for skill upgradation (v) frequent changes in placement (vi) vacant posts in remote areas (vii) lack of incentive structure (viii) low intensity of investment in extension (ix) reduction in non-salary expenditure and (x) inadequate information, management and evaluation system.

They further suggested several measures on institutional investment and management fronts to improve allotment and cost efficiency of extension systems *viz.* (i) deviating from past top-down to system oriented bottom-up approach (ii) strong government will to improve management (iii) improve the ability of extension workers (iv) incentive/merit based promotion (v) extension workers accountability (vi) feed back incorporating into performance evaluation (vii) higher plan allocation (viii) to sub-contract private sector for extension services and to attract private investment in extension directly (ix) extension diversification involving private sectors, NGOs, farmer groups, para-extension workers etc. and (x) well structured information and decision support system. Efficiency of these step are contingent upon government will and wherewithal to implement and ability of extension workers to avail them.

Van Den Ban and Hawkins (1996) reported that the constraints which decrease the effectiveness of agricultural extension organizations were; (i) appropriate technology is not available to extent to the farmers (ii) there are no effective linkage between extension organizations and research institutions (iii) field level personnel lack practical training in agricultural technology (iv) extension personnel lack training in extension methods and communication skills (v) extension personnel lack adequate transport facilities to reach farmers effectively (iv) extension personnel lack essential teaching aids, demonstration material and communication equipments and (vii) due to organizational problems, extension personnel have many other tasks besides extension work. They further identified the seven points to be considered by the administrators/managers of extension organizations in order to manage their organizations *viz*. (i) policy goals through communication and education or combined with other policy instruments (ii) effective communication between extension and research institutions (iii) to manage the planning of extension programme (iv) farmers participation in planning extension programme and in developing and testing of extension message (v) administrator's/manager's stimulation to develop organization through: staff capacity, clear goals, well trained staff, motivation and promotion (vi) good cooperation and effective communication and (vii) available resources.

2.5 Recent Reviewed Literature

Mangala Rai (2006) stated some of the salient features of the progress and achievements in agricultural research, education and extension and the initiatives taken by the council towards enhancing agricultural productivity. During the year, the council was able to established 128 new KVKs raising their number to 504 in the running year. ICAR have been operating partnership and linkages at the national and international level through memoranda of understanding (MOUs/work plant) for overall development of agrarian country like India.

Pawar (2006) mentioned the council has established a large network of Krishi Vigyan Kendra (KVKs) mainly for technology dissemination, refinement and front line demonstration of technologies. The farmers are benefited from technology demonstrations, training and extension activities undertaken by KVKs. We must address the issues of increasing unemployment, educated unemployment and massive underemployment in agriculture. Unless agriculture becomes economically rewarding, it will be highly difficult to attract or retain youth in farming.

Chitra Pathak *et al.* (2005) concluded that agriculture extension in the current scenario of a rapidly changing world, has been recognized as an essential mechanism for delivering knowledge (information) and advice as an input for modern farming. Besides performing the conventional task of transferring technology packages to the farming community. Hence, this era of new technological development training of both the extension functionaries and the farmers in the use and application of appropriate technologies should receive topmost priority. She further stated that the large network of front line ICAR – SAU extension system consists of 415 KVKs, 10 trainer's training centers (TTCs), 70 institution village linkage project centers and 44 agricultural technology information centers. Through this network training of 6.0 lakh farmers and farmwomen every year in agriculture and allied fields such as crop production, plant protection, livestock production and farm machinery and tools, and science is competed in the country, with limited resources at the disposal of SAUs and KVK in the sates, the effectiveness and productivity of training programmes are crucial for achieving the desired results.

Kalra and Mishra (2005) reported that we have still a very long way to go and a very active trial and error process has to be run a full course in order to evolve the desired institutional arrangements at grass root level. Scientists have to shoulder the main responsibly regarding arranging funds, planning and implementation of the project. Project could be successful in Krishi Vigyan Kendra (KVK) situated in the village and villagers take up the joint responsibility. Krishi Vigyan Kendra (KVKs) are the best institutes for effective transfer of technologies.

Singh and Padaria (2004) reported that superimposed on TOT approach was the T and V system with emphasis on the role of extension in technology transfer to encourage utilization or research results. The T and V approach was somewhat like a campaign approach an succeeded to some extent where the clients need and technology potential matched. The system was found to be too narrow in its approach and not suitable for small farms and rain-fed areas. IN between, the ICAR launched its programmes namely operation research project (ORP), Krishi Vigyan Kendra (KVK), and Lab to Land Programme (LLP) with the same objective in mind. All these programmes were later merged under the programmes of KVK. These programmes did receive some success but not upto the mark, particularly ensuring people's participation. The farming system research and extension programme was also started through KVKs. We have a strong research base of ICAR and university system and a large network of extension system. KVKs are the best institutes at grass root level for effective agricultural extension.

Lee and Kim (2004) showed that the E-linked KVKs can serve the problems for young and well educated people, working in agriculture it is easier to find employment elsewhere than for older people. Hence, a country with a rapid economic growth during last 40 years has 51 per cent of the farm operators were over 60 years and only 60 per cent were less than 40 years.

Swanson (2004) reported that the KVKs can train the extension personals for updating their knowledge with changing scenario of the agriculture at global level. There are large differences between countries in the opportunities farm families have to send their children to a vocational agriculture school. In India, a small proportion of these youngsters can get a course of one week of a farm science centre (KVK).

Sulaiman and Hall (2004) stated that it is indeed important that farmers are well informed about new production technologies, which may be profitable in their situation. These technologies are not only developed in government research institution but increasingly also in commercial companies selling inputs to farmer or marketing their products and by innovative farmers. Extension services should use all available information sources about new technologies. They further stated that the development of useful new production technologies requires partnerships between research institutes, companies selling inputs or marketing products and farmers. Unfortunately in many developing countries these partnerships do not exist or are weak. In a few Indian, districts these partnerships are promoted in a world bank project on agricultural technology management agency (ATMA). Krishi Vigyan Kendra has a prime importance in ATMA project.

Van den Ban (2004) concluded that in India thousands of agricultural graduates are trained for IT related job. They can get subsidy from the government for the necessary investments in ICT and make a living from selling information for which farmers feel a need. Agri-informatics and Bio-informatics at KVK is the advance step of this line.

Van den Ban (2004) showed that insufficient vocational agricultural education makes it difficult for Indian farmers to compete in the world market with farmers from other countries. It shows the importance of the farm science centres (KVKs) in Indian agriculture.

Vinayangam and Samanta (2004) stated that the organisations responsibility is to provide an congenial climate and satisfying environment in which people can gain overall satisfaction in their quality of work life (QWL) and help organisation to attain its objectives. It was also stated that if employee has a positive perception of QWL in the organization, he or she will strive to further improve the working conditions and increasing productivity. This is more so, and very much applicable to an organisation like 'Farm Science Centre" popularly known as 'Krishi Vigyan Kendra (KVK) established by Indian council of Agricultural Research (ICAR). The work environment of trainers in KVK is pivotal to make themselves fit into the system which generates motivation, commitment and dedication to transfer the technological information and skill to the farmers through training. The study was under taken to find out the perception of trainer's of KVK, working under four organizations namely ICAR, SAUs, State Department of Agriculture and NGOs.

Das (2003) reported that the Krishi Vigyan Kendra apart from having the internet access from the Pravara network also has the Agricultural Research Information System (ARIS) call set-up under the NAPT project of ICAR for access of information from the ICAR research centers and SAUs. The Krishi Vigyan Kendra at Babhaleshwar has internet network linking all the subject matter specialists (SMS) with all the villages in the network and to the internet as well. All the SMSs of the KVK frequently build up the information repository for agriculture extension at its server, which is available for all those connected to the wireless LAN in the Pravara Network. He also stated that a programme to establish a strong e-linkage with the KVKs of 200 districts and technology generating institutions has been prepared for availability of technology information. The objectives of the project are to establish a strong e-linkage with the KVKs and technology generating institutions like ICAR institutes and State Agricultural Universities for availability of technology information for the farmer at KVK level, to enable the KVKs for global access and to act as gateway for availability of such information at local level and to develop necessary human resource for handling cyber extension. The KVK will be equipped to enable them to convert the existing knowledge base into web-enabled electronic form to serve as information hub of the area.

Chapter 3

Theoretical Orientation

The main objective of this chapter is to present schematically the conceptual frame of reference used for this study. An attempt has been made to portray systematic framework, which would provide guidance in selection of variables. The chapter is presented in the following sections:

3.1 Extension management ability as a concept

3.2 Theoretical perspective

3.3 Independent variables

3.4 Derivation of hypotheses

3.5 The conceptual model

3.1 Extension Management Ability as a Concept

Management and administration have been used consciously as synonymous concepts in management literature. Management has been allowed almost as synonymous of administration.

The term 'Management" stems from the word "Manage" which is derived from French word "Manage" meaning 'House keeping'. In Latin literature, the term 'Management' is viewed to be derived from the word 'Manus" which means 'Hand" and it originally was referred almost exclusively to the handling and training of horses.

In general usage, the 'Management" identifies a special group of people whose job is to direct the efforts and activities of other people towards common objectives.

Irrespective of the nature of organization, there are some fundamental principles and methods which guide their effective functioning which in modern terminology we call it as 'management'.

According to Drucker and Fayol (1954), there is no much distinction between the terms 'management' and 'administration' and both are synonymous and used interchangeably. According to them, running a business or social institution or government office, person heading them requires specialized skill. In business, this specialized skill is called 'Management', while in government and social institution this specialized skill is called 'Administration'. But, a question is often asked whether the task of administration is the same as that of management. To probe further into the matter, we thus need to turn to common practices in the government institutions and industries. The distinction between the scope and connotation of the two terms administration and management is not qualitative. It lies in the emphasis. In case of administration, the emphasis, which tends to be wider, long ranged, diffused, comprehensive and more fundamental. While in case of management, the emphasis is more time bound, concreted, delimited, economically viable and more susceptible. The beneficiaries of the administration are at large in entire society, which are neither identified nor homogenous, while the beneficiaries of the management are a few *i.e.* the owners, the employees and indirectly the consumers and almost homogenous. Generally in administration, long term social costs and social gains occupy a central position, while in management the profit or monetary or economic return on efforts is almost necessarily in the forefront.

According to one school of thought, administration and management are different functions. This view is also supported by eminent writers like Oliver Sheldon, William R. Spriegel and G. W. Milward. The differences between them are administration is concerned with the functions as policy making, legislative, determinative and thinking relates to the apex and top level management and needs administrative ability rather than technical ability. While, management is concerned with the functions as policy implementation, executive governing and doing, relates to the middle and lower level management are required technical ability more than administrative ability. According to E. F. L. Brech (1972) management is a comprehensive generic term, which includes administration as a branch of it. The management function has been divided into two categories. The upper level management is usually called as administration management and the lower level of management is known as operative management. According to V. L.

Srivastava (1981), management is a social process entailing the execution of over all polity and plan *i.e.* functions of execution being also called proper management. While administration is that part of business management which is entrusted with the task of the determination of policy *i.e.* functions of direction being also called top management.

From the above views, it can be said that the management is a specialized set of functions/principles, which also includes the administration aspects. As the present study is limited to management aspects and viewed on management as a concept expressed by various thinkers are presented below.

According to Davis (1951), management is the function of executive leadership anywhere.

Peter Drucker (1954) defined management as a multipurpose organ that manages a business and manages manager's workers and works.

To, Niles (1956) good management or scientific management achieves social objectives with the best use of human and material, use of time and with satisfaction to the participants.

Vance (1956) defined management as simply the process of decision making and control over the actions of human beings for the purpose of attaining pre-determined goals.

Mee (1963) defined the term management as art of securing maximum results with minimum of efforts so as to secure maximum prosperity and happiness for both employer and employee and thereby giving best possible services to the public.

Terry (1968) defined management as a distinct process consisting of planning, organizing, actuating and controlling personnel to determine and accomplish the objectives by the use of people and resources.

Appley (1980) defined management as guiding human and physical resources into a dynamic and hard hitting organizational unit that attains its objectives to the satisfaction of those served and with a high degree of morale and sense of attainment on the part of those rendering the service.

According to Dasgupta (1969), management is the creation and control of technological and human environment or an organization in which human skills and capacities of individuals and groups find full scope for their effective use in order to accomplish the objectives for which an enterprise has been set up.

Morell (1969) viewed that management is that activity in an organization, which consists in deciding upon the means by which the goals are to be effectively reached.

Hodge and Johnson (1970) defined that management is the process of making decisions and issuing commands on behalf of an organization taking into consideration the complex of objectives, laminations and standards underlying the production and distributing of value required to satisfy members' needs.

Koontz and O'Donnell (1972) defined management as the art of getting things done through and with people in formally organized group, the art of creating an environment in such an organized group where people can perform as individuals and yet cooperate towards attainment of group goals and the art of optimizing efficiency in effectively reaching goals.

Management, according to McGreger (1971), is responsible for organizing the elements of productive enterprise *i.e.* money, materials, equipments, people etc. in the interest of economic ends.

According to Scanlan (1973), management involves the coordination and integration of all resources both human and technical to accomplish specific results. The five basic functions which have historically formed the core of studying management are planning, decision making organizing, directing and controlling.

Managerial process, according to Hellriegel and Slocum Jr. (1975) is concerned primarily with the design, planning and control of the organization's behaviour as well as the decision making which is required.

As view by Koontz *et al.* (1980), management is the task of managers to secure and utilize inputs to the enterprise, transforming them through the managerial functions of planning, organizing, staffing, leading and controlling with due considerations of external variables to produce outputs.

In these ways, many scientists have contributed to the meaning and concept of management in their own ways. Finally one can be define management as "the effective and efficient integration and coordination of resources to achieve desired objectives." The term extension management can be defined as an effective and efficient utilization and coordination of available resources to achieve the objectives of extension including Krishi Vigyan Kendras.

The term 'Ability' usually denotes a potential for performing some task. A person's ability to perform a task refers to the degree to which he possesses all the psychological attributes necessary for a high level of performance. Performance of an individual is the function of certain characteristics of the individual including his knowledge, skills, motivation, attribute and certain aspects of environmental situation (McGreger, 1967). This can be extended to programme coordinators of Krishi Vigyan Kendra also. Extension management ability of programme coordinator is thus, a function of his personal attributes, socio-psychological characteristics and the work environment.

To meet the needs of all the potential beneficiaries, programme coordinator has to develop better coordination of various extension systems and also has to share the responsibility of extension education with other agencies and organizations. All this calls for new policy initiatives, Creative management of the extension systems and organized efforts to upgrade the knowledge and skills of extension functionaries. Thus, a programme coordinator has to perform a variety of tasks related to extension management.

3.2 Theoretical Perspectives

The present belief of management or management theory or management science in the current situation can be understood in a better way if we have some idea about its evolution. This section is devoted to a critical review of some comprehensive and refractory hypotheses of management or management thought which is classified and presented into three phase's *viz.*, Pre-scientific Management Approaches, Early Management Approaches and Modern Management Approaches.

3.2.1 Pre-Scientific Management Approaches

3.2.1.1 Ancient Management Thoughts

Ancient Greek literature points to the existence of administrative apparatus like council, courts and boards of military personnel to ménage the affairs of the state.

Socrates, a famous philosopher was among the first to look upon management as something different from mere technical knowledge and skill.

The success of the Roman Empire depends on the ability of the rulers to organize their forces according to the scalar principle, neatly defining the area of authority and responsibility of each official.

3.2.1.2 Management of Roman Catholic Church

A set of well defined objective, an effective and efficient organization setup to achieve objectives, division of work based on specialization, employment of staff officers to aid and advice the line managers were some of the striking features of the management of Roman Catholic Church.

3.2.1.3 Managerial Practices by Military Organizations

Military organizations also contributed in their own simplistic way to the development of managerial practices though there was little area of theory in it. Even, so, their techniques of authority relationship between individuals and groups of individuals, direction, motivation and communication underwent considerable improvement over the years. Of late, military organizations have also experimented quite successfully, with the staff principle, according to which persons with specialized knowledge were appointed to aid and advise the field officers.

3.2.1.4 Management Theory by Cameralists

Contribution of management theory was also made by a group of German and Austrian administrators and thinkers, collectively called as the cameralists, who believed in organization of material wealth to make the state strong. The cameralists felt that to achieve this objective, the state needed to adopt the same techniques as an individual. Among the management principles formulated by the cameralists were; (a) job satisfaction (b) proper selection of personnel (c) division of work based on specialization and (d) effective control.

The above concepts were mostly related to the principles of specialization, selection of subordinates and their training and simplification of administration procedures. However, the various ideas stated by them have created awareness about managerial problems.

3.2.2 Early Management Approaches

3.2.2.1 Management by Custom School

Exponents of this school of management though emphasize the importance of a study of the experience of successful managers or a study of mistakes made in management. They claimed that such study would provide a better understanding of the most effective way of managing an enterprise.

3.2.2.2 The Decision Theory

Exponents of this school of though emphasize the decision making aspects of management.

3.2.2.3 Scientific Management Movement

Taylor (1915), due to his most significant contribution to the concept of scientific management was regarded as "The father of scientific management movement". He stated that scientific management in its essence consists of combination of great principles of management *viz.*, the development of true science, scientific selection of workers, intimate and friendly cooperation between the management and their workmen. It was based on a systematic study and analysis of various aspects of work involved with a view to finding out the best way of doing a thing, we call it scientific management

Gilbreth (1915) is regarded as "The father of motion study" and laid the entire foundation of our modern application of job simplification incentive wages plans, work standards and better work methods.

Gilbreth (1915)* emphasized the human factor and stressed physiological effects of fatigue.

Gantt (1919), contribution to scientific management includes task and bonus approach for remunerating workers, indicating a more humanitarian approach and emphasis on service rather than on profit.

Thus, scientific management emphasized faster and better methods of production and also increased out put and efficiency.

3.2.2.4 The Management Process School

The operational or management process fathered by Henry Fayol (1949) emphasized that management is an operational process and can best be studied through an analysis of the managerial functions. The process can basically be divided into five functions namely; planning, organizing, staffing, motivating and controlling although different authors use different words to describe these functions or divisions.

Koontz and O'Donnell (1972) divided the process into planning, organizing, staffing, directing and controlling. Terry (1968), preferred the word actuating to motivating. Gullick's (1925), formula of functions or process of management involves planning, organizing, staffing, directing, coordinating, reporting and budgeting (POSDCORB): Urwick and Gullick (1925) synthesized the ideas of various writers on the process management theory and indicated the probability of the science of management. Many principles of process theory of management have been applied with great success even today in all types of settings, though with some modifications.

3.2.2.5 Human Relation Movement

Elton Mayo (1932) "The founder of human relations movement", stated that the employee's morale had a great influence on productivity and the manager should treat them as social beings instead of economic beings or simply as cogs of wheel. The manager should understand group. The contributions of human relations remain and are being applied even today by managers.

3.2.3 Modern Management Approaches

3.2.3.1 Neo-Classical Theory

Neo-classical theory consists of two stream *viz.*, human relations and behavioral science approach. The important elements of this theory are; individual work groups, participative management, orientation, motivation, leadership and employee development. In this approach human factor and values are given more importance.

3.2.3.2 Social System Theory

Barnard Chester (1938) stated that the management is a social system and a system of cultural relationships. Cooperation and group spirit among members are necessary for achievement of organizational objectives. It is very much valuable to the practicing managers.

3.2.3.3 Quantitative Approach

Quantitative approach to management is also known as operational research in management science. This approach offered a systematic analysis and solution to many complex problems faced by management. For decision making on complex issues, mathematical and statistical tools are applied.

3.2.3.4 System Approach

Management is a system, which is a set of interconnected and interrelated elements/components/parts to achieve certain objectives. It has a number of sub-systems and classified into two groups *viz.* closed system and open system.

3.2.3.5 Situational Approach

According to this approach, different situations demand different solutions. There cannot be universal principles of management appropriate to all situations. This contingency or situational approach helps the management in finding solutions to the problems in specific situations.

3.2.3.6 Modern Management Thoughts

In the recent past, a number of persons have contributed to the development of management science. More important among them are Oliver Sheldon, Mooney and Railey, Chester Barnard, Lyndall, Uriwick and Peter Drucker. Peter Drucker outshines all other contemporary thinkers and he is considered as "The father of Modern Management Thoughts". Drucker considers management as a mere practice rather than a science. He advocates centralized control and decentralized structure.

Peter Drucker (1974) stated that every employee has crystallized and complex sets of needs, values, way of perceiving the world and his personality, which he brings with him to his work place. He is thus, not a raw mould. He tries to fit his psychological make up with the climate of the organization. The fact that his psychological characteristics influence his behaviour, suggests that his job behaviour and his motivation for job performance will also be influenced by these characteristics. That means, in hiring a worker, one always hires the whole man and this explains why the improvement of human effectiveness in work is the greatest opportunity for improvement of performance and output of the organization.

Vroom (1964) also emphasized that job performance is a result of interaction between worker's skill, ability and motivation.

3.2.4 Implication of the Theories

On the basis of the different theories discussed above, it could be inferred that the various functions/principles or practices of management for any organizations alongwith certain socio-economic forces operative at a given time provide a frame for individual to work effectively. The processes or practices of extension management and socio-economic forces may either enhance or restrict or at least lower the pace of the individual, the programme coordinator in the present study. The socio-economic dynamics enables him to be more able in extension management or restricting him to less extension management ability level.

These theories further lead to identify certain elements of management ability. These factors could be analyzed at various levels *viz.,* management functions/skills/practices, administrative management aspects, psychological aspects, motivational aspects, social system aspects, human relation aspects and many other related aspects *viz.,* planning and organizing the activities, proper direction of work, effective communication, control over work, human related aspects of staff, leading the entire programme of work, providing proper supervision to the subordinates, coordinating the programme activities in a well manner and take proper decision at various levels, were emerged as important indicators for extension management ability of programme coordinator of Krishi Vigyan Kendra and were felt appropriate for inclusion to understand extension management ability of programme coordinators.

The indicators as well as sub indicators of programme coordinator's extension management ability derived on the basis of different theories were subjected to judge's opinions after through considerations of management experts.

3.3 Independent Variables

The role of situation or environment is very crucial in understanding human action. Behaviour takes place in a situation and it has therefore, profound influence on the individual's action.

Sometimes, situations "present" the actor with certain goal to pursue; Rogers (1962) therefore confirmed the role of situation on determining an action of individuals.

It therefore, implies that the social system, in which individuals are members, has dominant effect on the behaviour of individuals. Members of social system characterized by traditional norms can be proven to resist change. This is a much widely accepted generalization.

Human action is dependent upon many factors. As conceptualized by Pearsons (1954), the action takes place in a situation consisting of social economic, physical and cultural factors and socio-economic dynamics operative at a given time. What action, an actor performed depends on how he orients to and evaluates the given situations *i.e.* the alternatives available and the expected gratification to be derived. The orientation and evaluation processes in turn depend on the personality and cultural systems. To be precise the action process is dependent upon personal, economic, social and psychological factors possessed by individual. Hence the analysis of action involves the concept and theories of various disciplines. He also mentioned that the selection of goals as well as means depends on such factors as cultural norms and values *i.e.* social, biological, capacities and accessibility to the physical and social environment.

On the basis of the previous discussion certain causative variables *viz.*, age, academic qualification, health, type of family, family size, native place, total working experience in KVK, experience as KVK head, total training received, availability of technical staff, span of control, interpersonal communication, organization climatic, organizational facility, infrastructural facility, extension staff facility, personality, perceived workload, attitude towards extension work, job involvement and job stress were selected to know their influence on extension management ability of programme coordinators of Krishi Vigyan Kendras.

3.3.1 Operationalization and Definition of the Independent Variables

3.3.1.1 Personal Variable

1. **Age**: It refers to number of chronological years completed by the respondent on the date of interview.

2. **Education:** It is formal education level of respondents.
3. **Health:** It is the physical and mental well-being of TOs as perceived by them in terms of very good health, Good health, Average health, poor health and Very poor health, it was measured as method used by Popat, (1991).
4. **Type of family:** It refers to type of family of TOs in terms of nuclear or joint type of family.
5. **Family size:** It refers to number of family members in the family of programme coordinator.
6. **Native Place:** In refers to the birth place of the programme coordinator. It is rural or urban background, surroundings, conditions and environment of the programme coordinators where he grew up.

3.3.1.2 Job Related Variables

7. **Experience as a Programme coordinator:** It is number of years of experience of the respondents as programme coordinators or in-charge of any KVK's for overall activities of KVKs.
8. **Total working experience:** It is total years of experience of working on the different cadres of TOs.
9. **Total training received:** It is total number of training received by TOs during service period in terms of short term training (up to 7 days), medium term training (8 to 15 days) and long term training (above 15 days), excluding M. Sc./Ph. D. study.

3.3.1.3 Organizational Variables

10. **Span of control:** The span of control is known as span of management. It is a human resources management term that refers to the number of subordinates a programme coordinator had to manage.
11. **Availability of Technical staff:** It refers to total number of technical personnel assisting to TOs in any KVKs to run overall activities of KVK. It was measured as per the method used by Patel (1990).
12. **Organizational climate:** It is overall atmosphere of KVKs as experienced by TOs. It was measured in terms of healthy,

good, average and adverse. It was measured as method developed by Talukadar, 1984.

13. **Organizational facilities:** It refers to physical facilities possessed by KVKs in terms of vehicle, audio-visual aids and other facilities. It was measured as method used by Patel (1993).

14. **Inter-personal communication:** It refers to degree or frequency of exchange of ideas or any information made by TOs with the higher and lower staff member. It was measured in terms of fortnight, monthly and more then monthly. It was measured by scale used by Popat (1991).

3.3.1.4 Socio-Psychological Variables

15. **Personality of TO:** It refers to introvert and extrovert qualities of TO, it was measured by scale developed by Woodworth (1979).

16. **Attitude towards extension work:** it refers to positive or negative feeling of TO towards extension work. It was measured by scale developed by Daftur (1986).

17. Workload: it refers to total workload possessed by TOs as perceived by them. It was measured in terms of very heavy, heavy, moderate and low. It was measured as method used by Patel (1990).

18. **Job involvement:** it refers to self-involvement of TO in administrative, extension, research and all other activities of KVKs. It was measured in terms of full, partial, occasional and no involvement in above activities it was measured as method developed by Lodalh and Kejner (1965).

19. **Job stress:** it refers degree of pressure and tension of work as perceived by TO during working period. It was measured in terms of the perception of TOs *viz.*, very high, medium and low. It was measured as method developed by Talukadar (1984).

20. **Constraints:** it refers to items of difficulties faced by the TOs in performing their activities.

3.4 Derivation of Hypotheses

Based on the theoretical framework and keeping objectives in view statement and importance of study, the following statistical hypotheses (null form: h0) were formulated as per the procedure recommended by kerlinger (1976).

$H0_1$: There will not be any relationship between personal, job oriented, socio-psychological and organizational character of programme coordinators of KVKs and their managerial ability.

$H0_2$: There will not be functional relation between managerial ability and various independent variables.

$H0_3$: There will not be any constraints faced by programme coordinators of KVKs in managing KVK activities.

3.5 The Conceptual Model

In the light of the above theoretical frame and the assumption derived there upon, a conceptual model delineating the relationship between independent variables and dependent variable has been proposed. The variables namely; age, academic qualification, family size, health, type of family, size of family, native place, total experience in KVK, experience as KVK head, total training received, span of control, interpersonal communication, organizational climate, infrastructure facility, technical staff facility, personality, perceived workload, attitude towards extension work, job involvement and job stress were taken as independent variables. The extension management ability was considered as dependent variable. Based on the above variables, the conceptual model and paradigm model were developed and it is shown in Fig. 5 and 6.

Chapter 4

Research Methodology

This chapter deals with research design, tools and techniques of a scientific investigation used for data collection in the light of objectives of the study. The selection of universe and sampling technique for investigation as well as devices used for data analysis are also explained in this chapter under following sub heads.

4.1 Plan of study

4.2 Local of the study

4.3 Sample size and selection of the respondents

4.4 Research Design

4.5 Tools for the study

4.6 Pre-testing of the Personal interview schedule cum questionnaire

4.7 Method of data collection

4.8 Selection of the variables

4.9 Measurement of dependent variable

4.10 Measurement of independent variables

4.11 Statistical framework for analysis of data

4.12 Derivation of research hypothesis

4.1 Plan of Study

The Krishi Vigyan Kendra (KVK) is an important agricultural extension project of the Indian Council of Agricultural Research (ICAR) to test and transfer agricultural technologies to bridge the gap between the production and productivity and to increase self-employment opportunities among farming communities. The training organizer as a head of KVK has to perform the role of extension manager to get the maximum output from the available resources.

The management ability of training organizer is important in achieving the maximum extension output for any KVK. To appreciate potential of KVKs, real understanding of the managerial ability of the training organizers is needed. However, the study on extension management ability of training organizers of KVKs was lacking. Only few studies have been conducted on role performance/ productivity/effectiveness of academicians, research personnel and extension workers. Hence, it was found worth to conduct a study on such an important aspect.

The idea was then discussed with research guide, teacher counselors and members of advisory committee as well as extension personnel. Finally, it was considered that the study of this nature would be fruitful and would provide new direction in the field of agricultural extension management. The outline of research work (ORW) was presented against panel of experts appointed by Yashwantrao Chavan Maharashtra Open University, Nashik, Maharashtra and got it approved for research investigation. This way it was decided to undertake the study on "Extension Management Ability of Training Organizers of Krishi Vigyan Kendras".

4.2 Local of the Study

The present study was carried out in all eight zones of Krishi Vigyan Kendras of India, where Krishi Vigyan Kendras are in operation in transfer of technology process. The ICAR is the apex body to conduct agricultural research and extension activities at the national level. KVK is one of the schemes of the council designed to provide skill oriented vocational training to the farmers, farmwomen, rural youths and field functionaries. The organizational structure has been suggested and developed by the council to maintain the scheme. It was therefore, decided to encompass the study on training organizers of about half of those KVKs which were in operation with full-fledged activities in the service of Indian farmers through effective agriculture extension services atleast with five years. At the time of investigation, about 320 well established KVKs were short listed. The KVKs in India are divided in eight zones for easy administration, thus purposively the representation of each zone was included in this study. All zones are headed by separate zonal coordinators at their head quarters/Zonal Coordinating Unit Office.

Zonal coordinators are in charge for monitoring, evaluating and controlling the KVKs of their concerned zones.

Thus, the present study was an attempt to evaluate and assess the extension management ability of the programme coordinators of KVKs of India, which is to some extent fresh work in this direction.

4.3 Sample Size and Selection of the Respondents

The present study was undertaken in all the eight zones of the country. The representation of all the states and all eight zones of India where, Krishi Vigyan Kendras are in operation in transfer of technology process was included for this study. A random sample of 12 to 20 programme coordinators of those KVKs which have completed atleast 5 years of working were selected from all the eight zones. Thus a random sample of 160 Training Organizers of KVKs was selected for this study. Out of selected total 160 programme coordinators of KVKs from the 28 States of India, 85, 56, 17, 1 and 1 KVKs were managed by SAUs, NGOs, ICAR, Deemed University and Open University, respectively. As far as state wise selection of programme coordinators of KVKs was concerned, from the Andaman and Nicobar-1, Andhra Pradesh-10, Arunachal Pradesh-1, Assam-4, Bihar-11, Chhattishgarh-6, Gujarat-12, Haryana-8, Himachal Pradesh 5, Jammu and Kashmir-1, Jharkhand-4, Karnataka-8, Kerala-10, Lakshdweep-1, Madhya Pradesh-6, Maharashtra-10, Manipur-1, Meghalaya-1, Mizoram-1, Nagaland-1, Orissa-10, Punjab-6, Rajasthan-11, Sikkim-1, Tamilnadu-8, Tripura-1, Uttar Pradesh-10, Uttaranchal-6 and from the West Bengal 5 programme coordinators of KVKs were selected for the study.

4.4 Research Design

As the study is concerned with the understanding the managerial ability of the programme coordinators or head of the KVKs, Ex-Post-Facto Research Design was used for this study. Kerlinger (1976) stated that the Ex-Post-Facto Research Design is worthy to apply when the independent variables have already acted upon. Ex-post facto research is systematic empirical investigation in which, the scientists do not have direct control on independent variables because they are inherently not manipulatable in nature. In this kind of research, researcher's control on the independent variable is very weak and in many cases no control is possible. Social science dose not afford a possibility of controlling the changes in the behavioural pattern of independent variables. These are usually affected by complex social phenomenon and one can only examine how the dependent variable is affected by such situations. In social and scientific research, there always exists a possibility of identifying dependent and independent variables either singly or in groups. Ex-Post-Facto Research is a good explanatory instrument of changes that take place in the dependent variables. If the set of independent variables is carefully selected and proper techniques adopted for analysis, meaningful insight can be gained into changing pattern of dependent variables.

4.5 Tools for the Study

The data were collected by personal interview as well as by mail questionnaire. Personal interview schedule cum questionnaire has been considered to be the most important tool and researcher can get most authentic first hand information. The interview schedule cum questionnaire was prepared by keeping in view the objectives of the study and was common for all the respondents. In formulating the questions and statements for the schedule, the investigator took opinion and guidance of the research guide, experts of the Department of Extension Education and Agricultural Statistics of B. A. College of Agriculture, Anand Agricultural University, Anand as well as experts of YCMOU, Nasik, Maharashtra.

The interview schedule cum questionnaire was divided into five parts. First part was containing some questions to collect personal and economic information, second part contained organizational information of KVKs. The third part was of socio-psychological information of programme coordinators. The fourth part was in relation to extension management ability of training organizers along with ten main indicators and 50 sub indicators of all main indicators. The fifth and final part of the interview schedule cum questionnaire was regarding the constraints faced by the programme coordinators and their suggestions to over come their constraints.

4.6 Pre-Testing of the Personal Interview Schedule-cum-Questionnaire

In order to test the administration ability of each item, question and statement and to remove ambiguity in measuring instrument, pre-testing of personal interview schedule cum questionnaire was carried out with 10 non-sample Training Organizers of zone 6, which comprises Gujarat and Rajasthan States. The training organizers were randomly selected from both of the states. The respondents were informed about the importance and purpose of the study. After gaining their confidence and developing rapport with them interview schedule was put forth to them. On the basis of answers given by these respondents and hurdles faced during interviewing, necessary changes were made after consultation with the research guide. The final approved interview schedule has been given in appendices.

4.7 Method of Data Collection

The data were collected by personal interview as well as mail questionnaire. The personal interview was carried out at the time of National Gathering of Krishi Vigyan Kendra of India at Rajasthan Agricultural University, Bikaner during 18-20 September-2005. Some of the responses were also collected from some of the programme coordinators by mail. While collecting data through personal interview, at the outset, the researcher introduced himself to the selected programme coordinators. The objectives of the study were explained with a view to facilitating correct responses. Assurance was given that the data collected would be utilized for the purpose of the study only. After establishing rapports, questions were asked one by one and their responses were recorded. Every possible cares were taken to maintain a favourable and unbiased opinion of the respondents. Similarly, while collecting responses through mail questionnaire; necessary cares were taken to make all the matters clear and understandable through written words. Before sending mail questionnaire, some times researcher also tried to contact some of the programme coordinators through telephone to clear the matter and purpose as well as to get better responses.

4.8 Selection of the Variables

The variables under study have been selected on the basis of extensive review of literature related to the subject and consultations with experts are as under. The major objective of this study was to measure the extension management ability of the training organizers of KVKs. Thus, the extension management ability of training organizer of KVKs was considered as dependent variable and overall characteristics of the programme coordinators were considered as independent variables.

4.8.1 Dependent Variable

1. Extension management ability.

4.8.2 Independent Variables

Personal Variable

1. Age
2. Educational qualification
3. Health status
4. Type of family
5. Size of family
6. Native place

Job Related Variables

7. Experience as a training organizer of KVKs
8. Total working experience in KVK
9. Total training received

Organizational Variables

10. Span of control
11. Availability of technical staff
12. Organizational climate
13. Organizational facilities
14. Inter-personal communication

Socio-Psychological Variables

15. Personality of the programme coordinators
16. Attitude towards extension work
17. Work load
18. Job involvement
19. Job stress

4.9 Measurement of Dependent Variable

4.9.1 Extension Management Ability of Training Organizers of KVK

The term Management is defined by different scientists in different words. "It is the ability of a person in using the techniques and skills in planning, scheduling, guiding, supervising and organizing the resources, monitoring and controlling of a research/ development or any other projects" (Venkateshwarlu, 1988). In the present study extension management ability of the programme coordinators of KVKs has been operationalized as the ability of the head to apply the basic principles of management in running the KVKs. The extension management ability of the programme coordinators of KVK was measured by using the scale developed by Patel (2001) with slight modifications.

Many scientists have contributed to the meaning and concept of management in their own way. Finally one can define management as the effective and efficient integration and coordination of resources to achieve desired objectives.

The term, extension management can be defined as an effective and efficient utilization and coordination of available resources to achieve the objectives of extension organizations including krishi vigyan kendra. The training organizer has to develop favourable coordination of various extension systems and also has to share the responsibility of extension education with other agencies and organizations. All this calls for new policy initiatives, creative management of the extension systems and all organized efforts to upgrade the knowledge and skills of extension functionaries. Thus, a training organizer has to perform a variety of tasks related to extension management.

To measure the extension management ability of the programme coordinators, the ten main indicators taken into consideration were; Planning, Organizing, Directing, Communicating, Controlling, Human relation, Leading, Supervising, Coordinating and Decision–making. The scale value for main and sub indicator were used as per the original scale. Total ten main indicators were measured by sub-indicators as seen in Appendix.

For measuring the extension management ability of programme coordinators of KVKs, the developed scale was applied to randomly selected respondents of selected KVKs. The responses were collected from the respondents on a three point continuum with categories *viz.*, 'Always', 'Sometimes', and 'Never' and the scores of 3, 2 and 1, were assigned, respectively. The formula used for calculating the extension management ability index was as under:

$$\text{EMAI} = \frac{\begin{matrix}\Sigma\,(\text{Score obtained for sub item} \times \\ \text{Weightage for sub item}) \times \\ \text{Scale value of main item}\end{matrix}}{\begin{matrix}\Sigma\,(\text{Maximum score for sub item} \times \\ \text{weightage for sub item}) \times \\ \text{Sclae value of main item}\end{matrix}} \times 100$$

The extension management ability index of all programme coordinators was calculated item wise. The final extension management ability index of programme coordinators was determined by averaging the index of all 10 sub-items of the respective programme coordinators. Then, the programme coordinators were classified into three categories on the basis of Mean and Standard Deviation, *viz.*, low (below mean – S.D.), medium (mean ± S.D.) and high (above mean + S.D.). Similarly, to classify programme coordinators in to three groups based on all the ten sub indicators also mean and standard deviation were used.

4.10 Measurement of Independent Variables

The details of measurement of independent variables are given below.

4.10.1 Age

Age of the programme coordinators was operationalised as number of completed years as on date of response. To quantify this variable one score to each year of age was given. The programme coordinators were classified into three groups, *viz.*, young (upto 35 years), middle (36 to 45 years) and old (above 45 years).

4.10.2 Educational Qualification

It refers to the programme coordinator's formal degree. The programme coordinators were classified into three groups as Doctoral degree holder (20 scores), Master's degree holder (17 score) and Bachelor's degree holder (15 score).

4.10.3 Health

The dictionary meaning of health is physical and mental well being, free from disease or a condition of body/mind. The health of programme coordinators was measured on the basis of their own perceptions. The programme coordinators were classified into five groups according to their self perception, *viz.*, very good health, good health, average health, poor health and very poor health and scores of 5, 4, 3, 2 and 1 were assigned to quantify this variable.

4.10.4 Type of Family

It refers to type of family of programme coordinators in terms of nuclear or joint type of family. The scores of 1 to nuclear type of family and 2 to joint type of family were assigned to compute this variable.

4.10.5 Size of the Family

Family size constitutes the total number of members of the family. One score was designed to each member of the family to quantify this variable one score to each member was given. The programme coordinators were grouped into three categories as small (below mean – S.D.), medium (mean ± S.D.) and big (above mean + S.D.).

4.10.6 Native Place

It refers to the birthplace of the training organizer. The programme coordinators were grouped into two groups *viz.*, "Rural " and "Urban" and scores of '2' and '1' were assigned, respectively to calculate this variable.

4.10.7 Experience as a Training Organizer of KVKs

It was operationalized as number of completed years of service as programme coordinators of KVK. The training organizers were classified into three categories *viz.*, low (below mean – S.D.), medium (mean ± S.D.) and high (above mean + S.D.) level of experience as programme coordinators of KVKs. One score was assigned to each year of experience to quantify this variable.

4.10.8 Total Working Experience in KVK

It refers to total years of working experience of programme coordinators on the different cadres in KVKs. It means length of service in KVK from the date of joining the service to the date of response was taken as total experience in KVK. One score was assigned to each year of experience to measure this variable. The training organizers were classified into three categories as low (below mean – S.D.), medium (mean ± S.D.) and high (above mean + S.D.) to working experience in KVK.

4.10.9 Total Training Received

Total training refers to the training received excluding M.Sc./ Ph.D. study by training organizers during service period. The programme coordinators were classified into two categories *viz.*, low (below mean) and high (above mean) trained. The scoring technique used to quantify this variable is given in appendix.

4.10.10 Span of Control

It is the total number of persons presently working in KVK under training organizer. The programme coordinators were classified in three categories *viz.*, low (below mean – S.D.), medium (mean ± S.D.) and high (above mean + S.D.). The scoring technique used to quantify this variable is given in appendix.

4.10.11 Availability of Technical Staff

It refers to total number of technical personals assisting programme coordinators in any KVKs to run overall activities of KVK. It was measured as per the method used by Patel (1990). It was measured on the three point continuum *viz.*, "more than requirement" (Excess), "As per requirement" (sufficient) and "less than requirement" (insufficient) and scores of 2, 3 and 1 were given, respectively. The same scoring method was used for each cadre *viz.*, technical, administrative and auxiliary staff. The programme coordinators were classified into three categories as low (below mean – S.D.), medium (mean ± S.D.) and high (above mean + S.D.).

4.10.12 Organizational Climate

It is overall atmosphere of KVK as experienced by programme coordinators. It was measured by using scale developed by Talukadar (1984). The responses were collected in terms of highly favourable, quite, some what and not at all favourable and the scores of 4, 3, 2 and 1, were assigned respectively to measure this variable. The training organizers were grouped in to three categories *viz.*, low (below mean – S.D.), medium (mean ± S.D.) and high (above mean + S.D.).

4.10.13 Organizational facilities

It refers to physical facilities possessed by KVKs in terms of vehicle, audiovisual aids and other infrastructural facilities for effective extension work of KVK. This variable was measured with the help of a developed questionnaire and responses were collected for each item in four point continuum *viz.*, available as per requirement, can be easily available from other, can available with some difficultly and can not be available and scores of 4, 3, 2, and 1 were assigned, respectively to measure this variable. The final score was received by summing scores of all the items. The programme coordinators were classified into three categories as low (below mean – S.D.), medium (mean ± S.D.) and high (above mean + S.D.).

4.10.14 Inter-personal Communication

It refers to degree or frequency of exchange of ideas or any information made by programme coordinators with the higher and lower staff members. It was measured by preparing questionnaire and responses were collected in five point's continuum in terms of contacts of programme coordinators with supporting staff. The responses were collected in terms of very frequently (every day), frequently (once in two days), sometimes (once on week), seldom (once in month or more period) and never and scores of 5, 4, 3, 2 and 1 were assigned, respectively. The final score was received by summing scores all the items. The programme coordinators were categorized into three groups as low (below mean – S.D.), medium (mean ± S.D.) and high (above mean + S.D.).

4.10.15 Personality of Training Organizer

There are two types of personality *viz.*, introvert and extrovert. The psychic energy of introvert programme coordinators is directed inward their thoughts, emotions and desires, whereas, the psychic energy of extrovert programme coordinators is directed outward to the social environment.

This variable was measured by adopting the scale developed by Woodworth (1979) with slight modification. The scale consisted of nine statements with a two point continuum *viz.*, 'yes and 'no' with score of 2 and 1 for positive and 1 and 2 for negative statements, respectively. The final score was obtained by summing scores of all the items. The programme coordinators were classified into two groups as introvert (up to Mean) and Extrovert (above mean).

4.10.16 Attitudes Towards Extension Work

It refers to degree of positive or negative feeling of programme coordinators associated with delegation of authority in KVK. It was measured by implementing the scale developed by Daftur (1986). The scale consisted of ten statements with a five point continuum as strongly agree, agree, undecided, disagree and strongly disagree with score of 5, 4, 3, 2 and 1 for positive and 1, 2, 3, 4 and 5 for negative statements, respectively. The final score was received by summing scores all the items. The programme coordinators were classified into three categories as low (below mean – S.D.), medium (mean ± S.D.) and high (above mean + S.D.).

4.10.17 Perceived Workload

It refers to the workload perceived by programme coordinators for their own job. The training organizers were classified on the basis of their perception of workload into three groups *viz.*, Heavy workload (three score), sufficient workload (two score) and light workload (one score).

4.10.18 Job Involvement

It refers to the degree to which a programme coordinators is involved psychologically with his work or degree of the importance of work in his total self image. It was measured by using a scale developed by Lodahal and Kejner (1965). The scale consisted of 20 statements with a five point continuum *viz.* strongly agree, agree, undecided, disagree and strongly disagree with the scores of 5, 4, 3, 2 and 1 for positive statements and 1, 2, 3, 4 and 5 for negative statements, respectively. The final score was gained by summing scores of all the items. The programme coordinators were classified into three categories *viz.*, low (below mean – S.D.), medium (mean ± S.D.) and high (above mean + S.D.).

4.10.19 Job Stress

It has been operationalized as the psychological and/or physical response state of the programme coordinators with regards to mental/physical stress upon him pertaining to his job. It was measured with the help of procedure adopted by Patel (2001). The responses were collected on a five point continuum *viz.*, strongly agree, agree, undecided, disagree and strongly disagree with the score of 5, 4, 3, 2 and 1 for positive and 1, 2, 3, 4 and 5 for negative statements, respectively. The final score was received by summing scores of all the items. The programme coordinators were grouped in to three categories as low (below mean – S.D.), medium (mean ± S.D.) and high (above mean + S.D.).

4.10.20 Constraints Faced by Programme Coordinators in Managing KVK and Suggestions Offered by them to Overcome Constraints

The constraints were obtained from all the respondent training organizers in managing KVK. To know their degree of importance, respondents were asked to give their responses in three point continuum *i.e.* very important, important and not important. The scores assigned were, two, one and zero for very important, important and not important responses, respectively. Finally, the mean score for each constraint was calculated for ranking them in terms of its importance. The same procedure was applied to identify and quantify suggestions.

4.11 Statistical Framework for Analysis of Data

The data collected were analyzed, classified and tabulated in order to make the findings meaningful for interpretation and drawing conclusions. For this, different statistical methods were used. Source of data was subjected to analyze in terms of percentage and frequency whenever necessary. At the other places, mean score and standard deviation were calculated. To find out the correlation between selected independent variables and extension management ability, the coefficient of correlation was computed. To predict the extension management ability by personal, job related, organization and socio-psychological variables, the stepwise regression analysis was done. The details regarding statistical framework is presented in this part.

The following statistical methods were used in the present study.

4.11.1 Frequency and Percentage

Simple comparisons were made on the basis of frequency and percentage.

4.11.2 Arithmetic Mean and Standard Deviation

These techniques were used for classification of respondents into different categories.

(*i*) The mean was obtained by dividing total score by the number of respondents.

$$\overline{X} = \frac{\Sigma Xi}{n}$$

where,

n = Number of observations

Xi Total score

(*ii*) The standard deviation was obtained by the square root of the average of the square deviation from mean.

$$S.D. = \sqrt{\left[\frac{\Sigma(Xi - \overline{X})^2}{n - 1}\right]}$$

where,

$\overline{X}$ = General mean

Xi = Observed values

S.D. = Standard deviation

n = Number of observations

4.11.3 Coefficient of Correlation

Coefficient of correlation was computed to find out the relationship between the independent variables and the dependent variable employing following formula. The Pearson's product moment method suggested by Chandel (1974) for computing correlation coefficient (r) was used to study the relationship between independent variables and dependent variable.

$$r = \frac{\Sigma XY - \frac{\Sigma X \Sigma Y}{n}}{\sqrt{\left[\Sigma X^2 - \frac{(\Sigma X)^2}{n}\right]\left[\Sigma Y^2 - \frac{(\Sigma Y)^2}{n}\right]}}$$

where,

r = correlation coefficient

X and Y are two variables

n = Number of observations

X = Independent variable

Y = Dependent variable

Σxy = Sum of product of the deviation of 'x' and 'y' from their mean

ΣX^2 = Sum of square of deviation of 'x' from their mean

ΣY^2 = Sum of square of deviation of 'y' from their mean

4.11.4 Stepwise Regression Analysis

The stepwise regression analysis was employed to predict the extension management ability by personal, job related, organizational and socio-psychological variables.

In the stepwise method, the regression analysis was started with regression of Y and X_1............X_k taking singly. The variable giving the greatest reduction in sum of squares of 'Y' was first selected. Then the bivariate regression in which X_1 appeared was worked out. The variate which gave the greatest additional reduction in sum of square after fitting X_1 was selected. All trivariate regressions that include both X_1 and X_2 were computed the analysis was continued till the last variate of which additional contribution was the least of all variates.

The prediction of equation was...........

$$Y = a + b_1x_1 + b_2x_2 + b_3x_3 + \dots\dots\dots + b_kx_k$$

where,

a = Intercept

b = Partial regression coefficient

x_1 ....x_k = Number of independent variables

After fitting the regression equation the't' value for partial regression coefficients was tested for their significance.

4.11.5 Standard Partial Regression Coefficient

In order to assign the rank to various selected impendent variable the standard partial regression coefficient was used. To calculate the same the following formula suggested by sendecor and cochran (1967) was used.

$$b' = \frac{\text{S.D. of individual item}}{\text{S.D. of total items}} \times \text{Partial 'b'}$$

where,

b′ = Standard partial regression coefficient

4.12 Derivation of Research Hypotheses

Keeping the objectives of the study in view the research hypothesis was formulated on different aspects of the study. While formulating the hypothesis nature or relationship between variables were determined on the basis of review of literature. The hypotheses were set up and presented in null form (h0).

$H0_1$: There will not be any relationship between personal, job oriented, socio-psychological and organizational character of programme coordinators of KVKs and their managerial ability.

$H0_2$ There will not be functional relation between managerial ability and various independent variables.

$H0_3$ There will not be any constraints faced by programme coordinators of KVKs in managing KVK activities.

Chapter 5

Findings and Discussion

The information pertinent to this study was collected from the programme coordinators through the interview schedules and questionnaires. The data thus, collected were classified, tabulated and analyzed in the light of the objectives of the study. The facts and findings derived after analyzing the information have been presented under the following main heads and discussed in succeeding pages:

5.1 Personal, job related, organizational and socio-psychological characteristics of the respondent programme coordinators

5.2 Extension management ability of the respondent programme coordinators

5.3 Relationship between Personal, job related, organizational and socio-psychological variables of the respondent programme coordinators and their extension management ability

5.4 The functional relationship between extension management ability and independent variables

5.5 Constraints faced by programme coordinators in managing KVK and suggestions offered by programme coordinators to overcome the constraints

5.1 Personal, Job Related,. Organizational and Socio-Psychological Characteristics of the Respondent Programme Coordinators

The extension management ability is influenced by different characteristics of the programme coordinators. It was beyond the scope of the present study to include all the characteristics of programme coordinators however, some important characteristics were selected and the findings have been presented in succeeding pages. The characteristics have been classified into four groups' *viz.*, personal, job related, organizational and socio-psychological.

5.1.1 Personal Characteristics

5.1.1.1 Age

Physical and psychological development of an individual is related to his age. It influences the interest and needs of an individual. Thus, age plays a crucial role in the behaviour of every individual. The information collected regarding the age of the respondents were grouped as (i) Young age group (up to 35 years); (ii) Middle age group (35 to 45years) and (iii) Old age group (above 45years). The data in respect of age of the respondents are presented in Table 1.

Table 1: Distribution of the programme coordinators according to their age

n = 160

Sl.No.	Age Group	Number	Per cent
1.	Young age group	15	9.37
2.	Middle age group	51	31.88
3.	Old age group	94	58.75
	Total	160	100.00

It is evident from the data presented in Table 1 that majority (58.75 per cent) of the programme coordinators were found in old age group, followed by 31.88 per cent with middle age group while only 9.37 per cent of the respondents were from young age group.

From the above result, it can be concluded that majority of the programme coordinators were from middle to old age group. The management of KVK is skillful work; person involved in such work has to deal with many people and situations. In addition to this, he has to do multifarious activities. To carry out such work effectively, certain skill, talent, expertise, proficiency and abilities are needed in any individuals. In such a situation handling of organization might have given to experienced persons. This might be the probable reason to have middle to old age among the majority of the programme coordinators of KVKs.

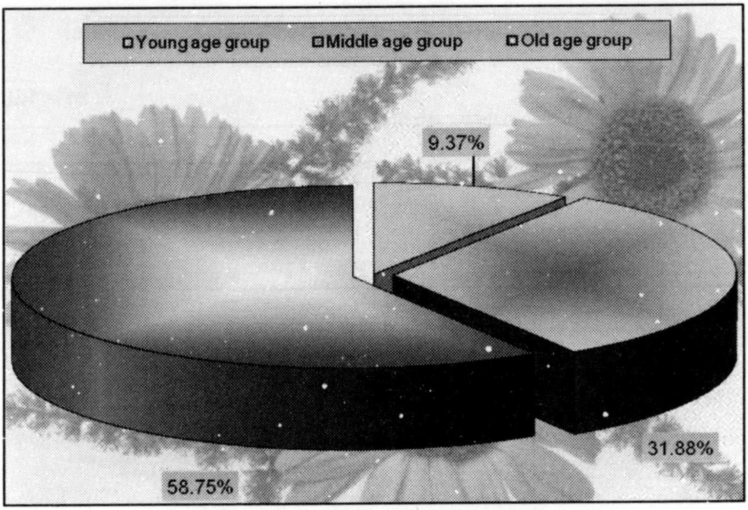

Figure 5: Distribution of the Programme Co-ordinator according to their age

Similar finding was reported by Patel (2001).

5.1.1.2 Educational Qualification

Education is generally considered as the process of producing desirable changes in human behaviour in terms of knowledge, skill and attitude. Education is valued as means of increasing knowledge and information. It facilitates learning. An individual's behaviour is influenced by his formal education. It leads him towards mental development and shapes view points. Keeping this in view, the level of education of the programme coordinators of KVKs was studied. The programme coordinators were categorized into three groups on the basis of educational qualifications such as Graduate, Postgraduate and Doctorate level. The data regarding educational qualification of programme coordinators are presented in Table 2.

The Table 2 gives the idea that majority (71.88 per cent) of the programme coordinators were educated up to doctorate level, followed by 19.37 per cent with master degree, while few of them (8.75 per cent) were educated up to bachelor degree only.

Table 2: Distribution of the respondents according to their education

n = 160

Sl.No.	*Category*	*Number*	*Per cent*
1.	Graduate degree	14	8.75
2.	Postgraduate degree	31	19.37
3.	Doctorate degree	115	71.88
	Total	160	100.00

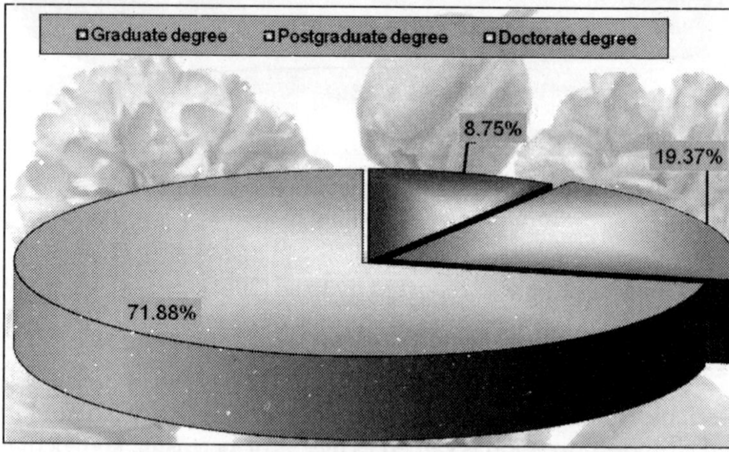

Figure 6: Distribution of the respondents according to their education

The result indicates that majority (71.88 per cent) of the programme coordinators working at KVKs were found with doctorate degree. As discussed earlier, person handling KVK needs ability to deal with many situations, activities and problems. For this reason, highly qualified persons might be preferred and appointed by the authority to run KVKs to achieve desired goals. This might be the reason to have doctorate degree among majority of the programme coordinators working at KVKs.

This finding was similar to the findings of Rani (1985), Samanta (1985), Padheria (1989) and Patel (2001).

5.1.1.3 Health Status

The dictionary meaning of health is physical and mental well being, free from disease. It is also defined as condition of body and/

or mind. It is believed that healthy person can do any work very effectively. Strong physical and mental health makes person able to think and do any work with his fullest involvement. Persons involved in any managerial work needs to have powerful physical and mental status. To understand effect of health on the extension management ability of programme coordinators, data were collected and presented in Table 3. The information on the health was collected and programme coordinators were classified into five categories according to self perception *viz.*, Very good health, Good health, Average health, Poor health and Very poor health.

Table 3: Distribution of the respondents according to their health

n = 160

Sl.No.	*Health*	*Number*	*Per cent*
1.	Very good health	34	21.25
2.	Good health	64	40.00
3.	Average health	58	36.25
4.	Poor health	2	1.25
5.	Very poor health	2	1.25
	Total	160	100.00

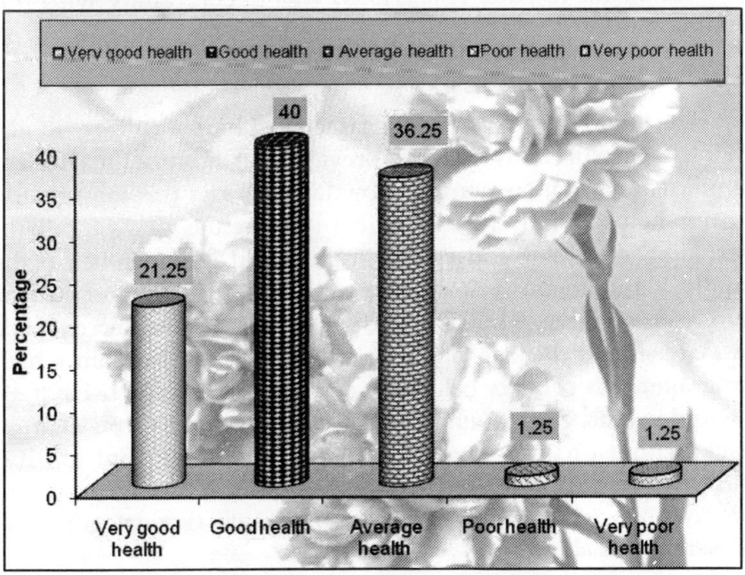

Figure 7: Distribution of the respondents according to their health

The data presented in Table 3 reflect that two fifth (40.00 per cent) of the programme coordinators realized that their health was good, followed by 36.25 per cent and 21.25 per cent of the programme coordinators perceived that they were having average health and very good health, respectively. Equal percentages (1.25 per cent) of the programme coordinators felt that status of their health was poor or very poor.

It can be summarized from the above result that majority (61.25 per cent) of the programme coordinators felt that status of their health was either good or very good. As discussed in previous table that programme coordinators have to do many activities, while doing mental and physical work at centre and in village areas he needs to maintain his physical and mental health, thus programme coordinators might have tried to maintain both the health. In addition to this, programme coordinators have to keep themselves busy and active while doing all such work, it is said that busy and active person's fitness remains always energetic and healthy. This might be probable reason to have good or very good level of perception about the health among majority of the programme coordinators working at KVKs.

The result of Patel *et al.* (1999) was in conformity with this finding.

5.1.1.4 Type of Family

The joint family system is character of Indian culture. In our country still joint family system is prevailing, at the same time nuclear families are also increasing day by day. It is said that joint family system helps an individual member to share the views and learn new knowledge, skill and attitude from the old aged members of the family. Simultaneously, it is assumed that because heads of nuclear family seem young or middle aged so there are more chances to take fast decision in nuclear family. Thus, type of family from which programme coordinator belongs to plays a big role on his behaviour. To understand consequence of programme coordinators' family background on their overall extension management ability, information was collected. As regards to the family type, the programme coordinators were classified into two categories as presented in Table 4.

Table 4: The respondents according to their type of family

n = 160

Sl.No.	*Type of Family*	*Number*	*Per cent*
1.	JOINT FAMILY	32	20.00
2.	Nuclear family	128	80.00
	Total	160	100.00

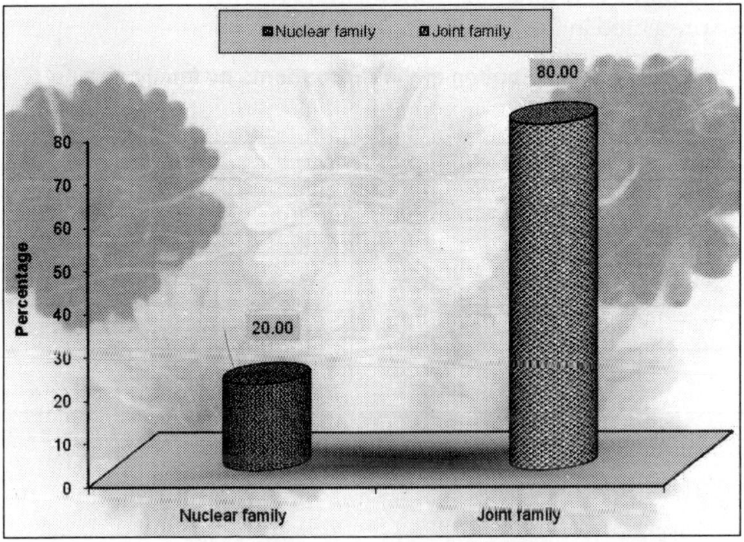

Figure 8: The respondents according to their type of family

The data presented in Table 4 indicate that majority (80.00 per cent) of the respondents had the nuclear type of family, while only 20.00 per cent of them had joint type of family.

Hence, it may be inferred from the above result that majority of the programme coordinators had nuclear type of family. The result presented in one of the subsequent tables says that majority of the programme coordinators were from urban background. It is observed that because of some limitations and needs urban people prefer to have nuclear family system. In addition to this, majority of the KVKs are located in rural or semi urban areas of the country. The programme coordinators working in KVKs have to stay near by the centers or in the premises of KVK; in this situation they might have preferred to live in nuclear family with wife and their children. This might be the reason to have nuclear family system, among majority of the programme coordinators.

5.1.1.5 Size of Family

The family size refers to the total number of persons live together in a family unit. It is also an important variable, which can play a role on the behaviour of an individual. Hence, the family size of the programme coordinators working in KVKs was studied and data are presented in the Table 5.

Table 5: Distribution of the respondents by family size

n = 160

Sl.No.	*Family Size*	*Number*	*Per cent*
1.	Up to 4 members	96	60.00
2.	5 to 8 members	50	32.00
3.	Above 8 members	14	08.00
	Total	160	100.00

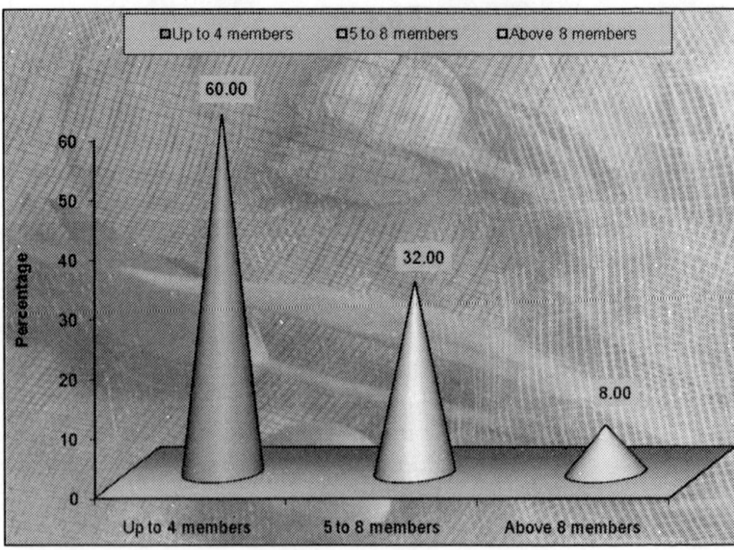

Figure 9: Distribution of the respondents by family size

The data presented in Table 5 reveal that majority (60.00 per cent) of the programme coordinators had up to four members in their family, while 32.00 per cent of them had five to eight members in their family and only 8.00 per cent had above eight members in their family.

From the above result, it can be concluded that majority (60.00 per cent) of the programme coordinators had up to four members in their family. The programme coordinators working in KVKs had nuclear type of family system; this might be the probable reason to have upto four family members among the majority of the programme coordinators of KVKs.

5.1.1.6 Native Place

The native place of programme coordinators of KVKs plays important role on their performance. It is assumed that persons, who have to work for the development of farmers should have knowledge about the life, social system, customs and psychology of the farmers. The programme coordinators of KVKs with rural background will have positive tendency towards farmers. To understand role of this factor on the performance of programme coordinators, information was collected and data with regards to the native place are presented in Table 6.

Table 6: Distribution of the respondents according to their native place

n = 160

Sl.No.	Native Place	Number	Per cent
1.	RURAL AREA	52	32.5
2.	Urban area	1.08	67.5
	Total	160	100.00

The data presented in Table 6 reveal that majority (67.50 per cent) of the programme coordinators of KVKs had urban background, whereas, 32.50 per cent of the programme coordinators of KVKs were from the rural area. The above finding indicates that majority of the persons working as programme coordinators in KVKs were not from rural background. To understand feelings, needs, problems, mindset, mentality and customs of farmers in a better way, authority

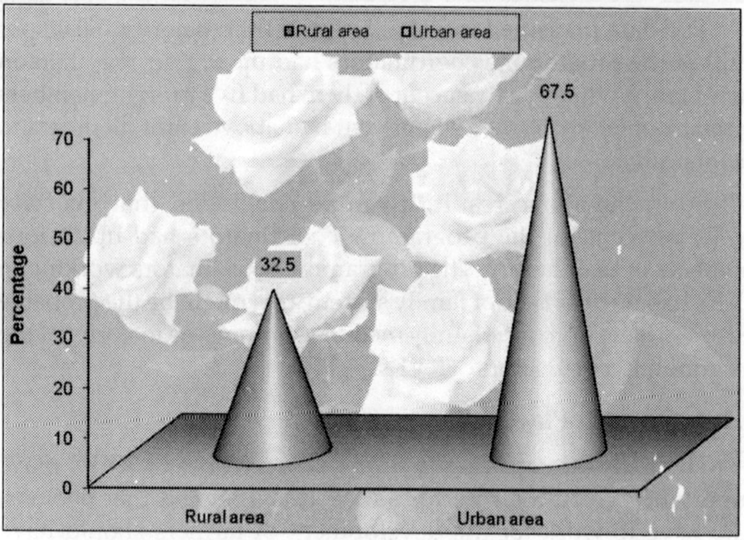

Figure 10: Distribution of the respondents according to their native place

should motivate to those who have rural background to work as programme coordinator of KVK. In other words, the result also reflects that persons from rural background either did not get chance or show deprived interest in working with rural people.

5.1.2 Job Related Variables

5.1.2.1 Experience as a Programme Coordinators in KVK

It is well known fact that man can learn from his own experiences. The experience has been considered as the best teacher. Man learns everyday new things from his day-to-day life and makes progress from trial and error through experiments. One can expect better performance on any work or job from those who have more years of experience to be a part of similar job or work. To understand effect of experience on the extension management ability on the performance of the programme coordinators of KVKs, information was collected and presented in Table 7.

Table 7: The respondents according to their experience of Programme Co-ordinator in KVK

n = 160

Sl.No.	Total Experience in KVK	Number	Per cent
1.	Low (below 3.98)	23	14.37
2.	Medium (in between 3.99 to 5.02)	118	73.76
3.	Wide (above 5.02)	19	11.87
	Total	160	100.00

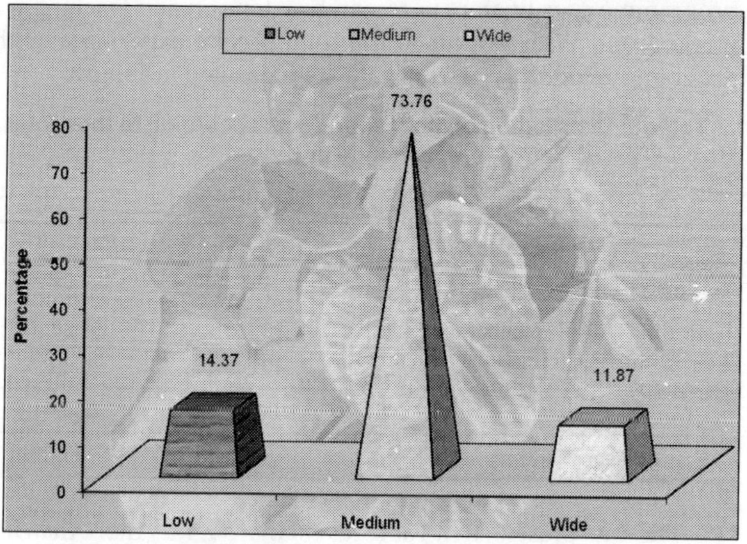

Figure 11: The respondents according to their experience of Programme Co-ordinator in KVK

The data seen in Table 7 reveal that majority (73.76 per cent) of the programme coordinators had medium level of total experience. The low and wide intensity of experience was observed among 14.37 per cent and 11.87 per cent of the programme coordinators, respectively. It is natural that establishment of KVK in India is in progressing stage and yet ICAR is trying to establish it in different states of India. It is therefore programme coordinators working in recently established KVK might have medium level of total experience in KVK.

The present finding was in contrast with the finding of Patel (2001).

5.1.2.2 Total Working Experience in KVK

The total experience of the programme coordinators indicates total year of working experience in the field of research, extension and teaching activities. Normally it is believed that varied experience of dissimilar activities, job or position gives opportunity to an individual to learn diversified activities, which leads person to improve his performance. Therefore, the total experience of programme coordinators in research, extension and teaching was worked out. With respect to the total service experience of the programme coordinators, the data are presented in Table 8.

Table 8: Distribution of the respondents according to their total experience in KVK

n = 160

Sl.No.	Total Experience	Number	Per cent
1.	Low (below 1.74)	37	23.12
2.	Medium (in between1.74 to 7.84)	83	51.88
3.	Wide (above 7.84)	40	25.00
	Total	160	100.00
	Mean = 4.79		S.D. = 3.05

The data presented in Table reveal that slightly more than half (51.88 per cent) of the programme coordinators had medium level of total experience in KVK on a different positions, whereas 25.00 per cent and 23.12 per cent of them had wide and low level of experience, respectively. As discussed in the previous table, it is obvious that founding of KVK in our country is in rising stage; however ICAR is attempting to launch at least one KVK in each district. As a result of this, programme coordinators working in the KVKs under this study might have medium level of working experience as a programme coordinator of KVK.

The finding of this study was in contrast with the findings reported by Patel *et al.* (1998)* and Patel (2001).

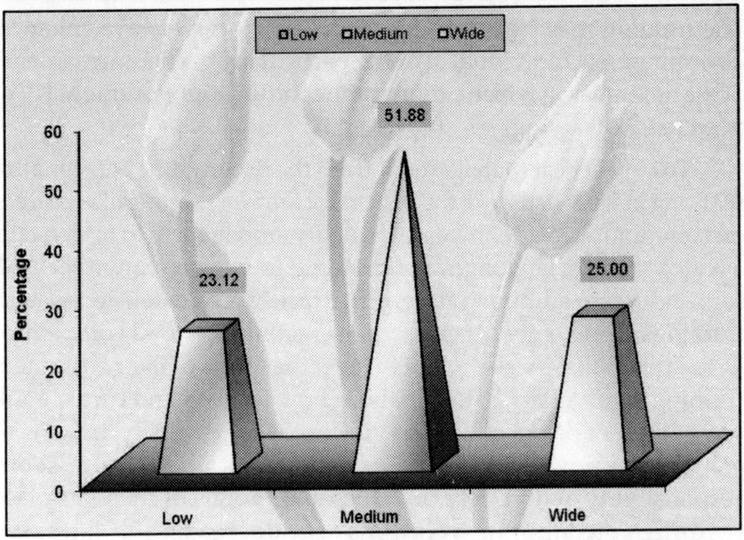

Figure 12: Distribution of the respondents according to their total experience in KVK

5.1.2.3 Training Received

As a brief review of terms, training involves an expert working with learners to convey certain areas of knowledge or skills to improve performance on current jobs. Development through training activities is a broad, ongoing multi-faceted set of activities to bring someone or an organization up to another threshold of performance, often to perform some job or new role in the future.

Training can be initiated for a variety of reasons for an employee or group of employees. It is preferred when a performance appraisal indicates to improve, to decide the standard of improvement, to expand an overall professional development programme, for succession planning to help an employee be eligible for a planned change in role in the organization, to test the operation of a new performance management system and to practice members of organization for specific subject matter.

Training improves the competency of the programme coordinators and keeps them informed on the latest developments. Training improves skill of trainee in doing his job efficiently. It also gives new techniques of work, orientation of forth coming practices.

The training may be in-service or pre-service; it always enhances performance of individual in doing particular job. There are varieties of the new areas in which programme coordinators working at KVKs need to take training.

To expect best management from the programme coordinators working in KVKs, some of the important areas of training like; training on communications to increase diversity among them to understand a wide variety of languages and customs, training on computer skills for conducting administrative, office tasks and e-extension system, training on customer service to understand increased competition in today's global market place to meet the needs of their customers, training on diversity to know about how farmers and rural people have different perspectives and views to value diversity, training on ethics to know increasing expectations about new social responsibility and variety of values and morals of the workplace, training on human relations to decrease job stresses, misunderstandings and conflict, training on quality initiatives to acquaint skill of total quality management, training on safety to know dangerous while working with heavy equipment, hazardous chemicals and repetitive activities and training on modern innovations to understand latest trends, are needed to achieve best results and determined objectives of KVK. Understanding importance of training in improving performance of the programme coordinators, information regarding total training received by them was collected and data are depicted in Table 9.

Table 9: Distribution of the respondents according to training received by respondents

n = 160

Sl.No.	*Training Received*	*Number*	*Per cent*
1.	Low (below 11.06)	122	76.25
2.	High (above 11.06)	38	23.75
	Total	160	100.00

It is observed from Table 9 that more than three-fourth (76.25 per cent) of the programme coordinators received less training, whereas slightly less than one fourth (23.75 per cent) of the programme coordinators received high training. It can be concluded

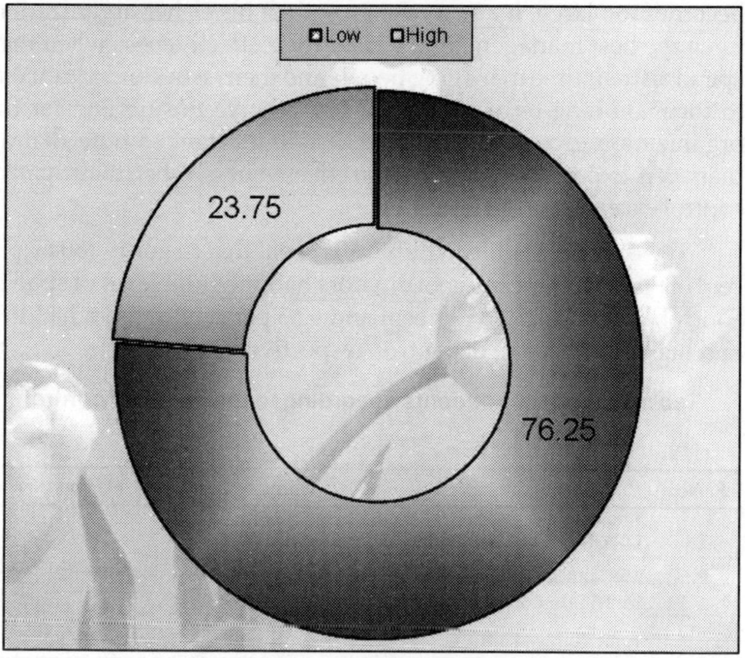

Figure 13: Distribution of the respondents according to training received by respondents

from the above data that majority of the programme coordinators received low training. Accepting important role of programme coordinators in doing effective activity of their respective KVKs for the overall development of farmers and rural India, they are needed to be enough competent through training in above listed areas to guide farmers to cope up with present scenario of WTO, diversified interests and diversified farming system.

5.1.3 Organizational Variables

5.1.3.1 Span of Control

The span of control is also known as span of management. It is a human resources management term that refers to the number of subordinates a manager or supervisor can effectively manage. It is important concept particularly for the effective management of any organizations. When an organization manager's span of control becomes too large, it affacts the growth of his or her organization. Even the best managers tend to lose their effectiveness when they spend all their time managing people and their issues and are unable to focus on long-term plans and competitive positioning for the organization growth as a whole. It is important to note that all managers experience a decrease in effectiveness, when their span of control exceeds the optimal level.

The data accessible in Table 10 reveal that majority (80.62 per cent) of the programme coordinators had medium level of span of control, whereas, 13.75 per cent and 5.63 per cent of them had low and high level of span of control, respectively.

Table 10: The respondents according to their span of control

n = 160

Sl.No.	Span of Control	Number	Per cent
1.	Low (below 8.18 score)	22	13.75
2.	Medium (in between 8.19 to 16.90 score)	129	80.62
3.	High (above16.90 score)	9	5.63
	Total	160	100.00
	Mean = 12.54		S.D. = 4.36

From the above information, it can be concluded that majority (80.62 per cent) of the programme coordinators had medium level of span of control. This might be due to staffing pattern decided for KVKs by the ICAR to work effectively.

Similar findings were reported by Patel (1993) and Patel (2001)

5.1.3.2 Staff Facility

It refers to the availability of technical/administrative/ auxiliary staff as per requirement perceived by programme coordinators working in the KVK. With respect to staff facility perceived by programme coordinators, the information is presented in Table 11.

It can be observed from the above table that exactly half (50.00 per cent) of the programme coordinators opined that the strength of staff provided them was medium, whereas, 26.25 per

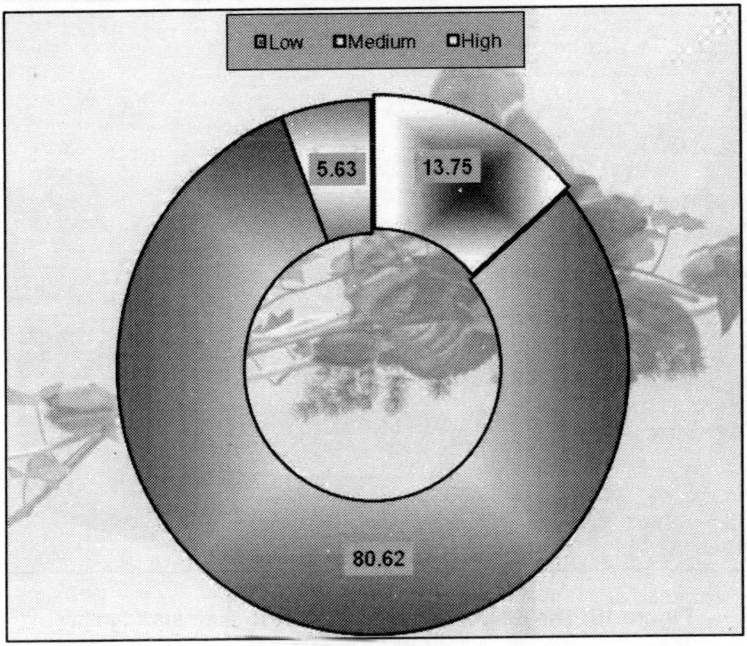

Figure 14: The respondents according to their span of control

cent of programme coordinators opined that existing staff strength was as per the requirement, whereas 23.75 per cent of them felt that insufficient staff was provided to their KVK.

Table 11: The respondents according to their staff facility

n = 160

Sl.No.	*Staff Facility*	*Number*	*Per cent*
1.	Low (below 3.93)	38	23.75
2.	Medium (in between 3.94 to 8.37 score)	80	50.00
3.	As per requirement (above 8.37 score)	42	26.25
	Total	160	100.00
	Mean = 6.15		S.D. = 2.22

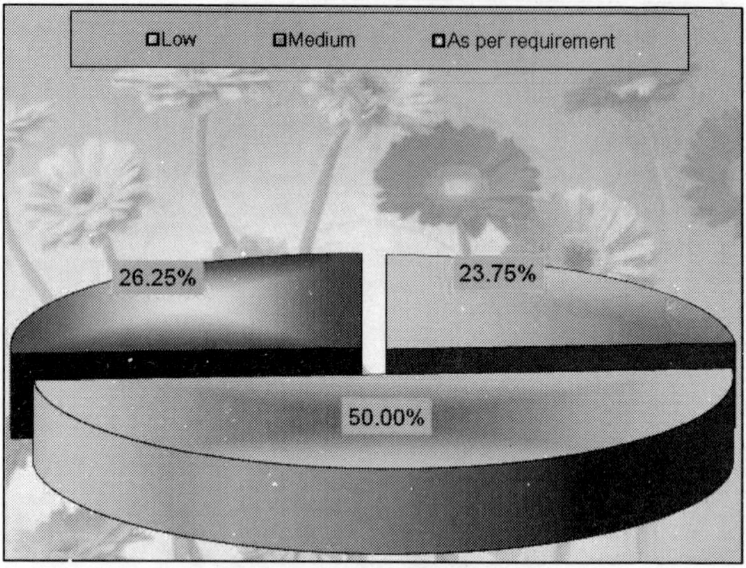

Figure 15: The respondents according to their staff facility

From the above result, it can be concluded that majority (76.25 per cent) of the programme coordinators perceived that the strength of staff provided in the KVK was medium or as per the requirement. The probable reason for this might be that nowadays ICAR is taking keen interest to strengthen farmers' development activity and giving good attention in recruitment of staff to carry out various extension activities of KVKs effectively. This way ICAR has tried to make KVKs well equipped up to a considerable level to provide vocational training to the practicing farmers.

5.1.3.3 Organizational Climate

The organizational climate is existing state of any organization as to grow continuously, manage with problems contribute to positive attitude, a feeling of belongingness and self-respect of working members that facilitate or hinder the effective functioning of the system. It is the environment, which helps in assisting and interfering with growth and bottom-line success. The organizational climate influences critical employee behaviours such as communication, problem solving, accountability, quality and performance. If organization wants to perform effectively, it is necessary to identify the key areas which are hindering performance, reducing effectiveness and which might generate unexpected impression. Organization becomes successful because of the quality of work, employees perform. When employees are cared for and the right environment is created where there are no barriers to performance, their true value to the organization can be fully realized.

Table 12: The respondents according to their organizational climate

n = 160

Sl.No.	*Organizational Climate*	*Number*	*Per cent*
1.	Low favourable (below 38.62)	26	16.25
2.	Medium favourable (in between 38.63 to 48.12)	111	69.38
3.	High favourable (above 48.12)	23	14.37
	Total	160	100.00
	Mean = 43.37		S.D. =4.75

The KVK is such an important rural and farmer development oriented organization, where favourable climate is needed. To

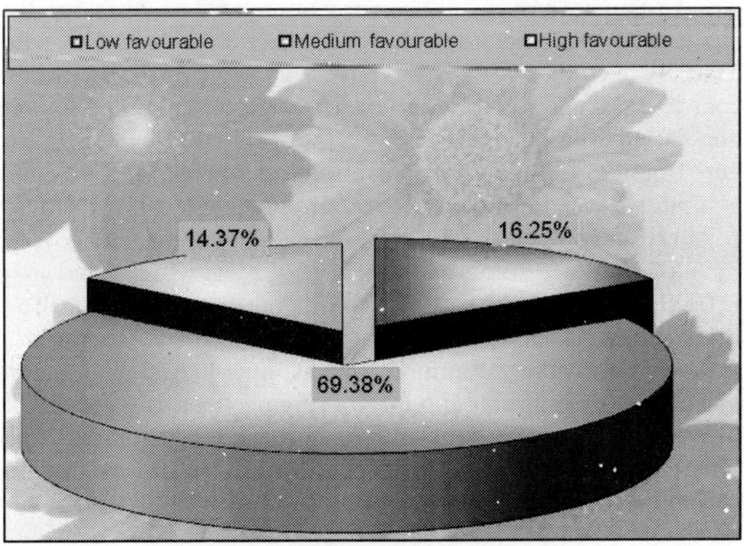

Figure 16: The respondents according to their organizational climate

understand programme coordinators' feelings about how employees of KVKs feel about their jobs, their supervisors, their peers, top management and many other factors affecting their individual productivity and collectively the ability of the KVKs to achieve its objectives, information on organizational climate of KVK as perceived by programme coordinators was collected and presented in Table 12.

It is evident from Table 12 that majority (69.38 per cent) of the programme coordinators working in KVKs had perceived medium level of favourable organizational climate, while 16.25 per cent and 14.37 per cent of the programme coordinators had low and high level of favourable perception for their organizational climate. From this result it can be said that majority of the programme coordinators had medium level of feelings about consciousness of their staff members' realization about sensitivity of each others happiness, clarity of goal and objectives of KVK, job satisfaction, confidence, trust, inter-personal discussions about the issues, degree of interest in taking initiatives and concern of doing work effectively.

This finding is in contrast to the finding reported by Patel (2001).

5.1.3.4 Organizational Facility

To achieve best result from any organizations, infrastructural facility plays a momentous role. KVK is such an organization, where a variety of works and activities have to carryout by the programme coordinators and his co-workers. The high-tech and employment generating training, publication and supply of literature, education and organization of demonstrations, e-extension, field visits, exhibitions, off campus training programmes, educational tour and celebration of special days to motivate rural farmers towards development are some of the key activities, which need many infrastructural facilities like land, laboratory, agricultural implements, vehicles, audio-visual aids, reading materials, furniture, stationary and other communication facilities like telephone, zerox machine, fax, computer and internet connections. It is proved that providing infrastructural facility is one of the key components to create workable and conducive atmosphere of any training, communication and academic organizations.

Table 13: The respondents according to their organizational facility

n = 160

Sl.No.	*Organizational Facility*	*Number*	*Per cent*
1.	Low (below 41.5)	08	05.00
2.	Medium (In between 41.6 to 57.54)	119	74.38
3.	High (above 57.54)	33	20.62
	Total	160	100.00
	Mean = 49.52		S.D. = 8.02

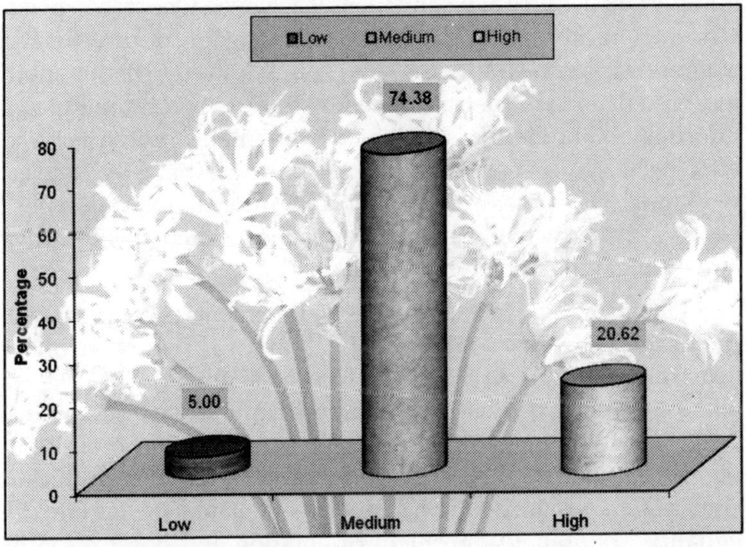

Figure 17: The respondents according to their organizational facility

It is the role of facility management to ensure that everything is available for the other staff in an organisation to do their work significantly. Facility may range from the small scale to the large scale facility. To understand status of infrastructure facility available at KVKs, information was collected and presented in Table 13.

The data in the same table reveal that majority (74.38 per cent) of the programme coordinators felt that existing organizational facility in their KVK was medium. On the other hand low and high level of organizational facility was perceived by 5.00 per cent and 20.62 per cent of the programme coordinators, respectively. It is important to note that ICAR is spending sizable budget to create all kinds of facility at KVKs.

5.1.3.5 Interpersonal Communication

Interpersonal communication is defined as communication that occurs between people who have known each other for some time. Importantly, these people view each other as unique individuals, not as people who are simply acting out social situations. There are many positive points of the interpersonal communication in organizations. The communication climate in any organization is a key determinant of its effectiveness. Interpersonal communication is important because of the functions it achieves. One reason persons working in any organization need to engage in interpersonal communication is so that they can gain knowledge about another individual. Social Penetration Theory says that we attempt to gain information about others so that we can interact with them more effectively. We can better predict how they will think, feel, and act if we know who they are. We gain this information passively by observing them actively, having others engage them or interactively and by engaging ourselves with them.

Organizations with supportive environments encourage worker participation, free and open exchange of information and constructive conflict resolution. In organizations with defensive climates, employees keep useful information to themselves, make only guarded statements and suffer from reduced morale. It is observed that employees are influenced by the communication climate in the organization. A supportive climate is having empathy, equality, spontaneity, problem orientation and description or a defensive climate is having evaluation, control, strategy, neutrality, superiority, and certainty.

In the overall development of any KVK, frequent contact of programme coordinator with their head, directors, officers of state agricultural university, state agricultural department and other, plays important role. Such contacts give opportunity to programme coordinators to discuss new problems, developmental issues, budget related matter, innovations, constraints and many other useful matters with other people.

Understanding the importance of interpersonal communication of programme coordinators for the effective management of KVK, data were collected and presented in Table 14. The data indicate that more than half (58.13 per cent) of the programme coordinators had medium level interpersonal communication, whereas, 23.75 per cent of the programme coordinators had low level of interpersonal communication. Only 18.12 per cent of the programme coordinators had high level of interpersonal communication.

Table 14: The respondents according to their interpersonal communication

n = 160

Sl.No.	*Interpersonal Communication*	*Number*	*Per cent*
1.	Low (below 22.25)	38	23.75
2.	Medium (in between 22.26 to28.89)	93	58.13
3.	High (above 28.89)	29	18.12
	Total	160	100.00
	Mean = 25.57		S.D. = 3.32

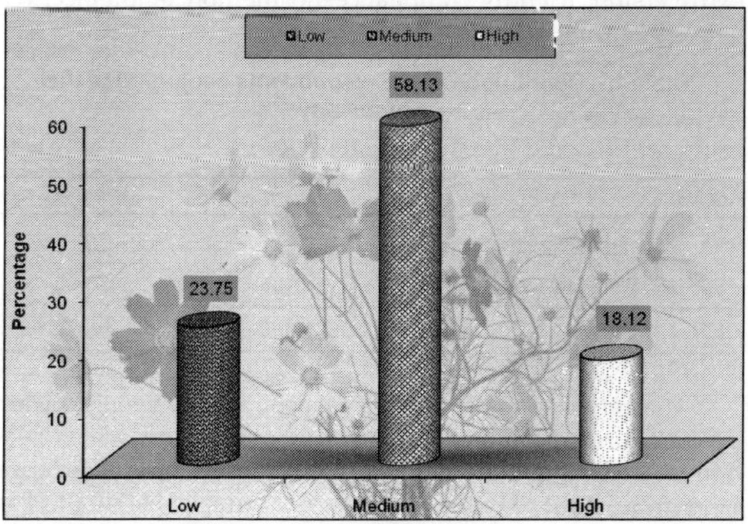

Figure 18: The respondents according to their interpersonal communication

From the above finding, it can be said that majority (81.88 per cent) of the programme coordinators had low to medium level of interpersonal communication. It is obvious that the programme coordinators have to do several activities to achieve mandates of KVK, thus, they might not have got enough time to contact repeatedly to the other officers. This might be the reason to have low to medium level of interpersonal communication among majority of the programme coordinators of KVKs.

5.1.4 Socio-Psychological Variables

5.1.4.1 Personality of the Training Organisers

The personality, in psychology is the patterns of behavior, thought and emotion unique to an individual and the ways they interact to help or hinder the adjustment of a person to other people and situations. The human mind could be divided into three significant components the id, the ego and the superego, which work together or come into conflict to shape personality. An individual's personality is developed through external stimuli. The psychologists have postulated five basic dimensions of personality; introversion-extroversion, friendly compliance–hostile noncompliance, will, neuroticism, and openness to experience.

Table 15: Distribution of the respondents according to their personality

n = 160

Sl.No.	*Personality*	*Number*	*Per cent*
1.	Introvert (below 13.65)	37	23.12
2.	Extrovert (above 13.65)	123	76.88
	Total	160	100.00

The person who has to work to inspire many people towards positive direction should have mature personality, which is not dominated by infantile pleasure drives. He should be characterized by a strong need to repeat certain acts or rituals. He should have enough confidence and feelings of social responsibility. The programme coordinators working at KVKs should have specific qualities to work effectively with rural people. Understanding this, personality in terms of likeness of programme coordinators to work with group, motivation, self confidence, friendly attitude, likeness

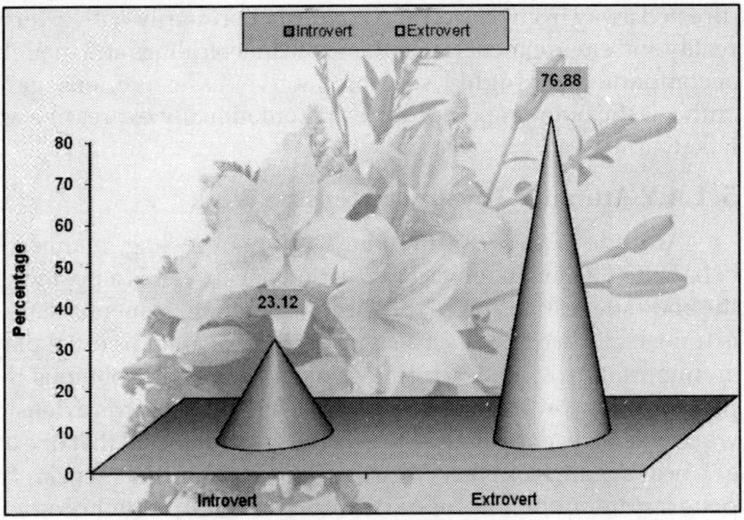

Figure 19: Distribution of the respondents according to their personality

to lead social gathering was thought to study, information regarding this is presented in Table 15.

It can be observed from the Table 15 that majority of the programme coordinators exhibited extrovert personality. The probable reason for emerging as extrovert category might be that the programme coordinators in the present study have long experience in extension management as well as extension work. Thus, it can be said that majority of the programme coordinators were gregarious and unreserved persons. This kind of personalities have act, state or habit of being predominantly concerned with and obtaining gratification from what is outside the self. Such qualities are helpful to carryout effective training, communication and educational and motivational work. Thus, it was interesting to note that majority of the programme coordinators working in KVKs had interest to know in what is happening around them, open mind and often talkative, mentality to compare their own opinions with the opinions of others, likeness of action and initiative and nature to make new friends or new group, interest in new people.

Extrovert Personality helps person to do extension work very effectively because of combination of traits and attitudes which are directed away from oneself and concerned primarily with external reality and environment rather than with inner feelings and thoughts accompanied by highly sociable, active, assertive, energetic, enthusiastic outgoing, impulsive and emotionally expressive and talkative.

5.1.4.2 Attitudes Towards Extension Work

Attitude is a way of thinking, acting, or feeling, manner, or behavior of a person toward a situation or cause. It is a position of the body suggesting an action, purpose, emotion, or mental state. It is universally accepted fact that an attitude of an individual plays an important role in determining his behavior. It is obvious that programme coordinators with positive feelings towards extension work will have more interest to do such work. It is said that one can get best result from any work when he loves that work. The programme coordinator with more mental, psychological, intellectual and intellectual attachment with extension activity will always lead his organizations towards grand success. Hence, the programme coordinators' attitude towards extension work was also included in the focus of present study. The distribution of the respondents according to their attitude towards extension work is presented in Table 16.

Table 16: The respondents according to attitudes towards extension work

n = 160

Sl.No.	Attitudes towards Extension Work	Number	Per cent
1.	Less favorable (below 32.95)	34	21.50
2.	Favorable (in between 32.95 to 45.25)	100	62.50
3.	Highly favorable (above 45.25)	26	16.00
	Total	160	100.00
	Mean = 39.10		S.D. = 6.15

It is evident from Table16 that majority (62.50 per cent) of the programme coordinators had favorable attitudes towards extension work, followed by 21.50 per cent and 16.00 per cent of them had less

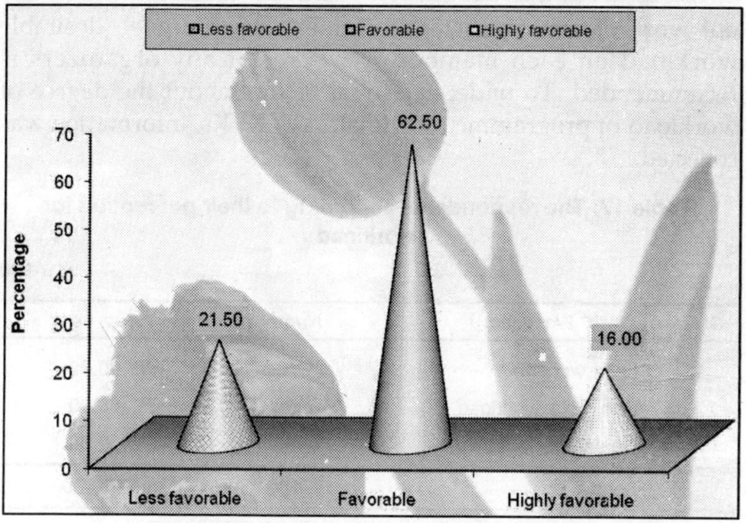

Figure 20: The respondents according to attitudes towards extension work

favorable and highly favorable attitudes towards extension work, respectively.

It can be concluded that great majority (78.75 per cent) of the programme coordinators of KVKs had favorable to highly favorable attitudes towards extension work. The reason behind this might be that the programme coordinators must have joined KVK with the feelings to do better service of nation and farmers understanding the main aim of the KVK. It is incredible to make a note of such mental status among the majority of the programme coordinators of KVKs. Favorable attitude towards extension work among programme coordinators helps to create constructive, encouraging and sympathetic impression of KVK among rural farmers to connect themselves with KVK activities.

5.1.4.3 Workload

Workload is the amount of work carried by programme coordinators or assigned to a worker or position. The workload plays important role on the performance of any manager or organizers. Heavy workload effects performance of any manager in undesirable manner, on other hand light workload shows more than required number of human resource in organization. It is therefore, to carry out work of any organizers to achieve desired goal, desirable workload on each member or worker of any organizers is recommended. To understand real picture about the degree of workload of programme coordinators of KVKs, information was collected.

Table 17: The respondents according to their perception for workload

n = 160

Sl.No.	Workload	Number	Per cent
1.	Light workload	1	00.62
2.	Desirable workload	96	60.00
3.	Heavy workload	63	39.38
	Total	160	100.00

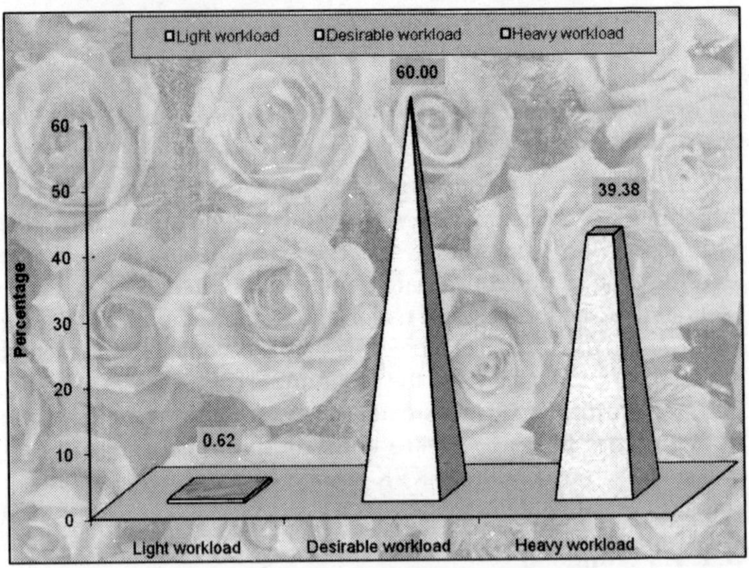

Figure 21: The respondents according to their perception for workload

The data in Table 17 revealed that majority (60.00 per cent) of the programme coordinators had perceived desirable workload, whereas, 39.38 per cent of the programme coordinators perceived heavy workload. Only 00.62 per cent of the programme coordinators had perception of light workload. Thus, It can be concluded that majority (60.00 per cent) of the programme coordinators of KVKs had perception of desirable level of workload. The result says that authority had tried to give desired level of workload to the programme coordinators of KVKs in an effort to anticipate best performance from them.

This finding is in contrast to the findings reported by Patel (1993) and Patel (2001).

5.1.4.4 Job Involvement

The simple meaning of job involvement is the action of involving with job, profession, work or duty. In this study, job involvement of the programme coordinators was measured in terms of degree to which he is involved psychologically with his work or what he thinks about significance of work. It is obvious that, for excellent result at the end of any work, person who has to carry out that work should have rational, intellectual, logical and constructive mental status of being involved in work. The favourable feelings towards assigned duty, job, aspiration to do work, willingness to bear responsibility, concern to the job and likeness of job are some of the important mental preconditions enquired among the programme coordinators to lead KVK towards desired directions. Understanding importance of this, information on the job involvement of programme coordinators was collected. The data regarding this are depicted in Table 18.

Table 18: The respondents according to their job involvement

n = 160

Sl.No.	*Job Involvement*	*Number*	*Per cent*
1.	Less favourable (below 64.90)	18	11.25
2.	Reasonable (in between 64.90 to 81.14)	128	80.00
3.	Highly favourable (above 81.14)	14	8.75
	Total	160	100.00
	Mean = 64.90		S.D. = 81.14

The data in Table 18 indicate that majority (80.00 per cent) of the programme coordinators were reasonable, followed by 11.25 per

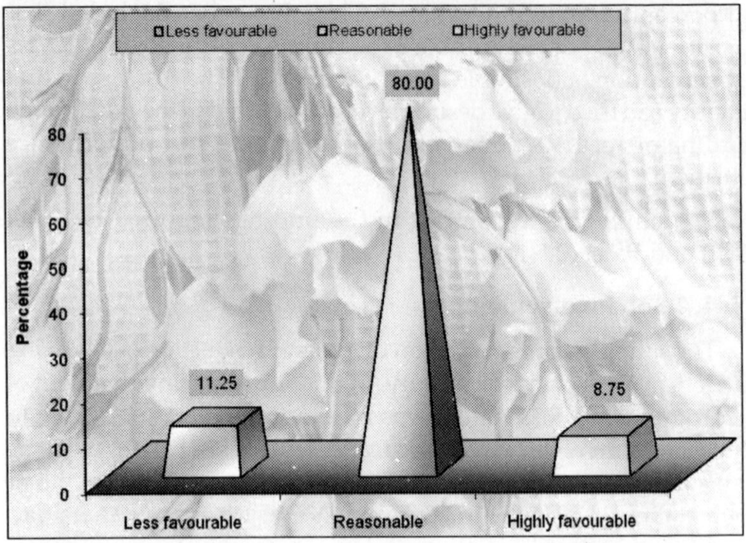

Figure 22: The respondents according to their job involvement

cent with less and 8.75 per cent with high favourable towards their job.

It can be included that majority of the respondents exhibited reasonable level of job involvement. The probable reason for the reasonable level of job involvement may be that the programme coordinators in the present study possessed favourable level of attitudes towards extension work.

This finding is in contrast to the findings reported by Patel (1993) and Patel (2001).

5.1.4.5 Job Stress

The job stress has been operationalized as reactions of programme coordinator with regards to their psychological and/or physical stress pertaining to their job. A stressful situation occurs when imbalance is observed between demands imposed upon programme coordinators by the circumstances and their ability to cope with demands. It is clear that KVK is not only human resource development centre but many other administrative, research related and technical works are also expected functions of it. Thus, the person who is leading KVK should have physically and mentally powerful mindset to do work in any dam situation. He should be enough strong, prompt, tension free, matured, accommodating and able to do any kind of work effectively with whatever available workforce he has. Understanding importance of ability to deal with stress situation in the leader of KVKs, fact about level of job stress of programme coordinators was studied.

From the data presented in Table 19, it is observed that majority (73.12 per cent) of the programme coordinators experienced medium level of job stress, whereas, 13.75 per cent with low level of job stress and 13.13 per cent of them had perception of high level of job stress.

Table 19: Distribution of the respondents according to their job stress

n = 160

Sl.No.	*Job Stress*	*Number*	*Per cent*
1.	Low (below 50.70)	22	13.75
2.	Medium (In between 50.70 to 68.82)	117	73.12
3.	High (above 68.82)	21	13.13
	Total	160	100.00
	Mean = 59.76		S.D. = 9.06

It can be inferred from the above results that majority of the programme coordinators had experienced medium level of job stress. The majority of the programme coordinators under study had four to five years of working experience, it is said that experience makes man perfect with job and same time it makes him able to tolerate various kinds of stress. An experienced person learns by trial and error to tackle different situations while working. This might be the reason to have low to medium level of perception about the job stress among majority of the programme coordinators.

This finding is in line with the findings of Talukdar (1984) and Patel (2001).

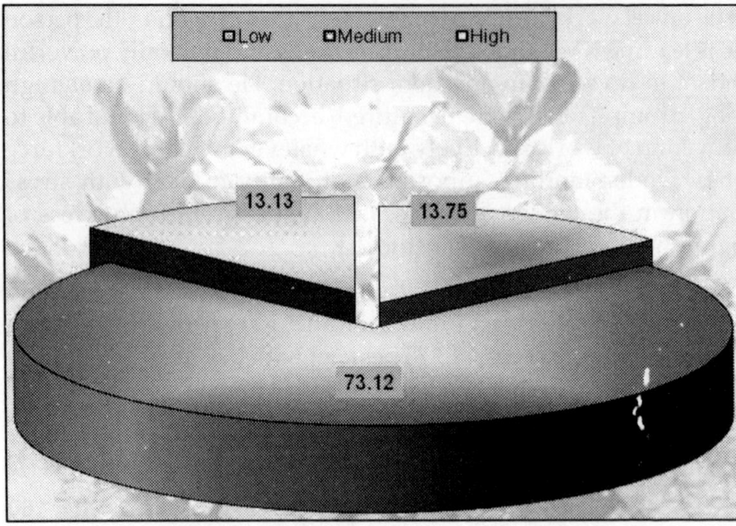

Figure 23: Distribution of the respondents according to their job stress

5.2 Extension Management Ability of the Respondent Programme Coordinators

To measure the extension management ability of Programme coordinators, ten indicators were studied *viz.*, planning, organizing, directing, communicating, controlling, human relation, leading, supervising, coordinating and decision making. The responses were obtained in three point continuums *viz.*, "always", "some time" and "never". The mean and standard deviation were worked out and based on mean and S.D., the programme coordinators were categorized in three categories such as low (Mean – S.D.), medium (Mean ± S.D.) and high (Mean + S.D.) in relation to all of the ten main indicators. The details were presented in following tables.

5.2.1 Ability of Planning

The Planning is one of the important indicators of extension management ability of the programme coordinators. It shows scheduling style implemented by programme coordinators to determine the objectives, training needs, process of transferring technologies, utilization of funds, extension facilities and monitoring and evaluation work. With a view to understanding ability of planning activity as one of the indicators of extension management ability of programme coordinators, information was collected and presented in Table 20.

Table 20: Distribution of the respondents according to ability of planning

n = 160

Sl.No.	*Ability of Planning*	*Number*	*Per cent*
1.	LOW	24	15.00
2.	Medium	54	33.75
3.	High	82	51.25
	Total	160	100.00
	Mean= 22.52		S.D. = 1.98

The data presented in Table indicated that in case of planning, slightly more than half (51.25 per cent) of the programme coordinators had high level of planning ability, followed by 33.75 per cent and 15.00 per cent of them with medium and low level of planning ability, respectively. It is concluded that (85.00 per cent) of the programme coordinator found with medium to high level of planning ability. Proper guideline, systematic direction and continous supervision by concerned officers and ICAR might have made them able with medium to high level of planning ability.

5.2.2 Ability of Organizing

The ability to organize different functions is one of the important indicators of extension management ability requires in any programme coordinators to arrange various activities on a regular basis. To call regular meeting to provide work opportunity according to specializations and to invite suggestions from subordinates as well as to divide activity into groups and to delegate authority according to responsibility are important functions which require proper ability to organize functions. With a view to understanding this ability as one of the indicators of extension management ability of programme coordinators, information was collected and presented in Table 21.

Table 21: The respondents according to ability of organizing

n = 160

Sl.No.	Ability of Organizing	Number	Per cent
1.	Low	24	15.00
2.	Medium	68	42.50
3.	High	68	42.50
	Total	160	100.00
	Mean = 16.70		S.D. = 1.54

The data in Table indicate that an equal number of (42.50 per cent) of the programme coordinators had medium and high level of organizing ability, followed by only 15.00 per cent of them had low level of organizing capacities. It can be clearly stated that majority (85.00 per cent) of the programme coordinators had medium to high level of organizing capacity in relation to extension management in their respective KVKs. The probable reason might be that programme coordinators have to carryout series of activities throughout the year; as a result of this they might have understood the importance to systematize various activities for their effective implementations.

5.2.3 Ability of Directing

This variable was measured by knowing degree of frequency of programme coordinators in directing their subordinates about different functions. This is an important activity; programme coordinators have to carry out for effective functioning of the KVK work. In order to know this ability as one of the indicators of extension management ability of programme coordinators, information was collected and presented in Table 22.

Table 22: The respondents according to ability of directing

n = 160

Sl.No.	Ability of Directing	Number	Per cent
1.	Low	15	09.38
2.	Medium	22	13.75
3.	High	123	76.87
	Total	160	100.00
	Mean = 11.65		S.D. = 0.75

It can be seen that majority of the (76.87 per cent) programme coordinators had high level of ability of directing their subordinates, followed by 13.75 and 9.38 per cent of them were with medium and low level of ability of directing their subordinates for efficient extension management activities of the KVKs. The results indicates that majority of the programme coordinators took high level of interest to give clear instruction of work, to give due credit to a person, to appreciate good work done by their subordinates and to give continuous information.

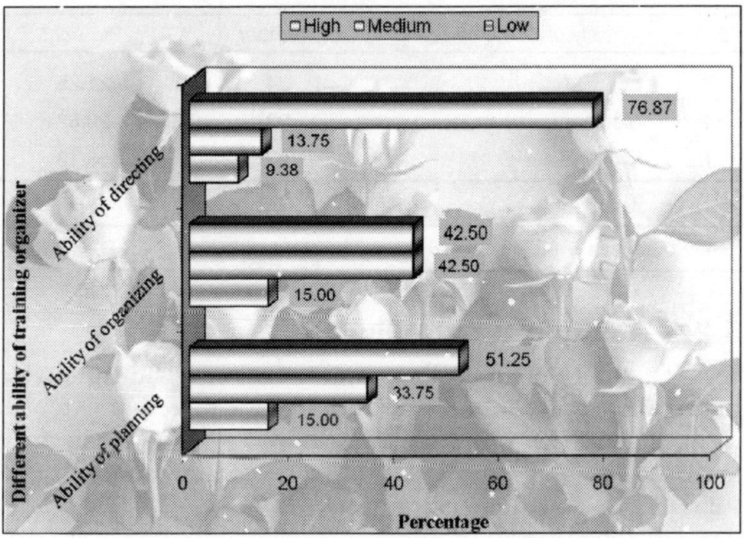

Figure 24: The respondents as per their planning, organizing and directing ability

5.2.4 Ability of Communicating

The ability of communicating subordinates is also an important indicator, which requires in programme coordinators for effective extension management. Understanding importance of this ability as one of the indicators of extension management of programme coordinators, information was collected and presented in Table 23.

It is seen from the Table that majority of the programme coordinators (58.13 per cent) had high level of ability of communicating, followed by 21.25 per cent of them had low and

20.62 per cent of them had medium level of ability of communicating, respectively. Thus, it can be concluded from the result that majority of the programme coordinators had high level of ability of communicating. The result indicates that majority of the programme coordinators had taken constant interest to give timely instructions, to listen subordinates, to select proper medium for instructions and to receive feed back

Table 23: The respondents according to ability of communicating

n = 160

Sl.No.	Ability of Communicating	Number	Per cent
1.	Low	34	21.25
2.	Medium	33	20.62
3.	High	93	58.13
	Total	160	100.00
	Mean = 11.31		S.D. = 1.15

5.2.5 Ability of Controlling

The ability of controlling is also an important sign, which requires in programme coordinators for effective extension management. This variable was measured by knowing degree of frequency of programme coordinators' command of ling their subordinates and different functions.

Table 24: The respondents according to ability of controlling

n = 160

Sl.No.	Ability of Controlling	Number	Per cent
1.	LOW	45	28.12
2.	Medium	38	23.76
3.	High	77	48.12
	Total	160	100.00
	Mean = 13.63		S.D. = 1.51

The data presented in Table stated that slightly less than half (48.12 per cent) of the programme coordinators had high level of controlling power, followed by 28.12 per cent and 23.76 per cent of them had low and medium level of controlling power, respectively. It is seen that majority (71.88 per cent) of the programme coordinator realized medium to high level of controlling power for effective functioning of extension activities in terms of monitoring the work, accurate and proper style of visits, determining standard for work, evaluating the work and measuring performance of staff.

5.2.6 Ability to Maintain Human Relation

The ability to maintain human relation is very important quality; programme coordinators need to maintain effective extension management. This variable was measured by knowing degree of frequency of programme coordinators' touch with persons to carry out functions of KVK effectively. Accepting importance of this ability as one of the indicators of extension management in programme coordinators, information was collected and presented in Table 25.

Table 25: The respondents as per ability to maintain human relation

n = 160

Sl.No.	*Ability of Human Relation*	*Number*	*Per cent*
1.	Low	13	08.13
2.	Medium	70	43.75
3.	High	77	48.12
	Total	160	100.00
	Mean = 17.01		S.D. = 1.32

The data revealed that great majority (91.87per cent) of the programme coordinators had medium to high level of capacity to maintain human relation, followed by only 8.13 per cent of them had low level of human relation making capacity. It is obvious from the table that majority of the programme coordinators had medium to high level of interest to make personal contacts, deal fairly, pay attention in development of juniors, adopt participative approach, understand others' behaviour and to offer good co-operation.

5.2.7 Ability of Leading

The ability of leading is significant feature; programme coordinators need to uphold for effective extension management.

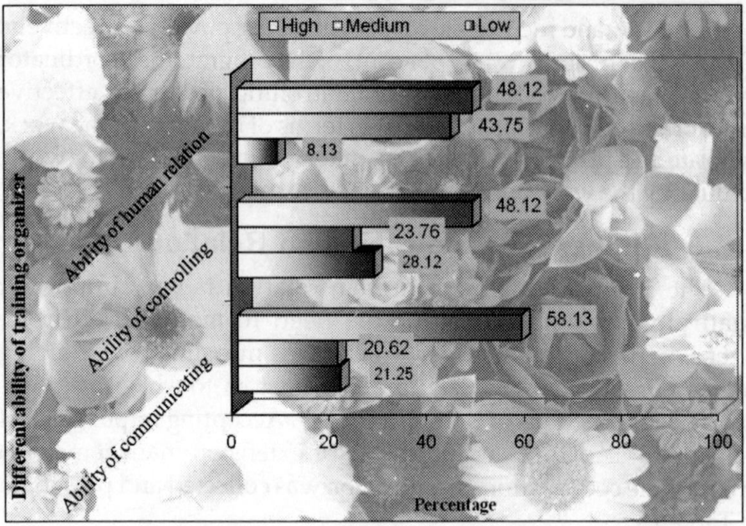

Figure 25: The respondents as per their communicating, controlling and human relation making ability

This variable was measured by knowing degree of programme coordinators' lead taking behaviour to perform the functions of KVK successfully. Considering importance of this aptitude as one of the indicators of extension management among programme coordinators, information was collected and presented in Table 26.

Table 26: The respondents as per their ability of leading

n = 160

Sl.No.	*Ability of Leading*	*Number*	*Per cent*
1.	Low	29	18.12
2.	Medium	32	20.00
3.	High	99	61.88
	Total	160	100
	Mean = 14.29		S.D. = 1.16

The perusal of the data presented in the table indicates that majority (61.88 per cent) of the programme coordinators had high level of feeling of lead taking behaviour, followed by 20.00 per cent and 18.12 per cent of them had medium and low level of feeling of lead taking behaviour, respectively. It can be concluded from the results that great majority (81.88 per cent) of the programme coordinators realized that they had medium to high degree capacity to take lead in different functions like; to promote team work, take initiative task, apply technical competency, have a sense of responsibility and to inspire subordinates.

5.2.8 Ability of Supervising

The ability of supervising is another considerable trait; programme coordinators require for upholding effective extension management. This quality was measured by knowing degree of programme coordinators' administering behaviour to do various functions of KVK successfully. Considering essence of this talent as one of the indicators of extension management among programme coordinators, data were collected and presented in Table 27.

Table 27: The respondents as per their ability of supervising

n = 160

Sl.No.	*Ability of Supervising*	*Number*	*Per cent*
1.	Low	13	8.12
2.	Medium	33	20.62
3.	High	114	71.26
	Total	160	100.00
	Mean = 8.60		S.D. = 0.71

The data presented in Table 27 clearly indicate that majority (71.26 per cent) of the programme coordinators indicated high level of supervising capacities, followed by 20.62 per cent and 8.12 per cent of them had medium and low level of supervising capacities, respectively. Thus, it can be said that great majority (91.88 per cent) of the programme coordinator had medium to high level of ability to regulate various functions by providing constant extensive counseling, focusing on quality work and giving valuable suggestions. Due to frequent involvement in various kinds of activity by the programme coordinators of KVK, they might have established such ability in them.

5.2.9 Ability of Coordinating

The ability of coordinating is one more substantial attribute; programme coordinators require for keeping efficient extension management. This element was calculated by insightful degree of programme coordinators' behaviour to organize, synchronize, harmonize, match up and bring together various on-campus and off-campus functions of KVK fruitfully. Considering core of this talent as one of the indicators of extension management ability of programme coordinators, facts were collected and presented in Table 28.

Table 28: The respondents as per their ability of coordinating

n = 160

Sl.No.	*Ability of Coordinating*	*Number*	*Per cent*
1	Low	43	26.87
2	Medium	17	10.63
3	High	100	62.50
	Total	160	100.00
	Mean = 11.21		S.D. = 1.15

It can be seen that majority of the programme coordinators were found with high level of coordinating ability (62.50 per cent), followed by 26.87 per cent of them with low and only 10.63 per cent of them with medium level of ability of coordination with concerned personnel in the mandatory working of the KVK. Thus, it was interesting to note that majority of the programme coordinators had high level of ability to establish linkages, integrate the extension work, establish mutual trust and get co-operation from outsides. The kind of works, programme coordinators have to do as a manager of KVK indicates that need to deal with internal and external personnel and agencies, while working with them, they might have realized importance to establish such ability. This might be the reason for above finding.

5.2.10 Ability of Decision Making

The ability of taking right decision at the right time is an added considerable characteristic; programme coordinators want for keeping competent extension management. This ingredient was studied by knowing degree of programme coordinators' behaviour to take different judgments to carry out various on-campus and off-campus activities of KVK at right time by proper ways. Taking into consideration, role of this capacity as one of the indicators of extension management ability of programme coordinators, information was collected and presented in Table 29.

Table 29: The respondents as per their ability of decision making

n = 160

Sl.No.	*Ability of Decision Making*	*Number*	*Per cent*
1.	Low	23	14.38
2.	Medium	65	40.62
3.	High	72	45.00
	Total	160	100.00
	Mean = 13.99		S.D. = 1.33

The data presented in Table indicate that 45.00 per cent of the programme coordinators had high level of decision making capacity, followed by 40.62 per cent of them with medium and only 14.38 per

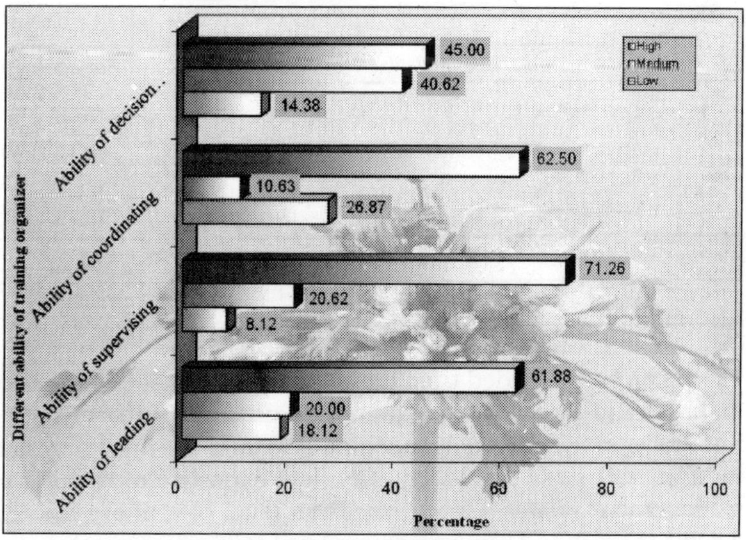

Figure 26: The respondents as per their leading, supervising, coordinating and decision making ability

cent of them with low level of decision making capacities, respectively. It can be stated that great majority (85.62 per cent) of the programme coordinators had medium to high level capacity to take timely decision, find out facts for decision, adopt participative decision style, taking unbiased decision for implementation.

5.2.11 Overall Extension Management Ability

The overall extension management ability of Programme coordinators was measured by combining all the indicators including planning, organizing, directing, communicating, controlling, human relation, leading, supervising, coordinating and decision making. The detail is presented in Table 30.

It can be seen that majority (68.12 per cent) of the programme coordinators were observed with the medium level of overall extension management ability, while 18.75 per cent with high and 13.13 per cent with low level of overall extension management ability.

Thus, the extension management ability of the respondents was predominantly medium.

Table 30: The respondents according to extension management ability

n = 160

Sl.No.	Extension Management Ability	Number	Per cent
1.	Low (below 73.33)	21	13.13
2.	Medium (in between 73.33 to 86.13)	109	68.12
3.	High (above 86.13)	30	18.75
	Total	160	100.00
	Mean = 79.73		S.D. = 6.40

It can be concluded from the above finding that the extension management ability of programme coordinators of Krishi Vigyan Kendra was predominately medium. This might be due to constant involvement in so many activities, less exposure of training on management related aspects, medium level of attitude towards extension work and having medium level of infrastructural facility.

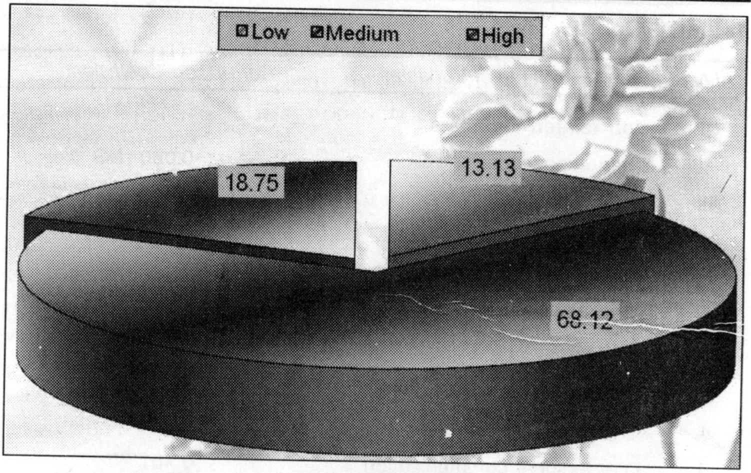

Figure 27: The respondents according to overall extension management ability

5.3 Relationship between Personal, Job Related, Organizational and Socio-Psychological Variables of the Respondent Training Organizers and their Extension Management Ability

Table 31: Relationship between the characteristics of the respondents and their extension management ability

n = 160

Sl.No.	*Name of Variables*	*Correlation Coefficient*	
I	**Personal variables**		
1.	Age	-0.362	**
2.	Educational qualification	0.238	*
3.	Health	0.365	**
4.	Type of family	-0.011	NS
5.	Size of family	0.020	NS
6.	Native place	0.384	**

Contd...

Table 31–Contd...

Sl.No.	Name of Variables	Correlation Coefficient
II	**Job related variables**	
1.	Total experience as TO	-0.060 NS
2.	Total experience in KVK	-0.013 NS
3.	Training received	-0.019 NS
III	**Organizational variables**	
1.	Span of control	0.144 NS
2.	Staff facility	-0.120 NS
3.	Organizational climate	0.524 **
4.	Organizational facility	0.537 **
5.	Interpersonal communication	0.401 **
IV	**Socio-psychological variables**	
1.	Personality	0.675 **
2.	Attitudes towards extension work	0.622 **
3.	Workload	0.023 NS
4.	Job involvement	-0.047 NS
5.	Job stress	-0.158 *

* Significant at 0.05 level of probability.

** Highly significant at 0.01 level of probability.

To ascertain the relationship between personal, job related, organizational and socio-psychological variables of the respondents and their extension management ability, correlation coefficient was worked out. On the basis of operational measures developed for the variables, null hypothesis were stated for testing the relationship and their level of significance are given Table 31.

5.3.1 Age and Extension Management Ability

In respect of age, the data presented in Table 31 indicate that the correlation coefficient (r= -0.362) was negative and significant with extension management ability of programme coordinators. The result says that management ability was observed better among those programme coordinators who had young age. In other words young programme coordinators were observed superior in planning, organizing, directing, communicating, controlling, leading,

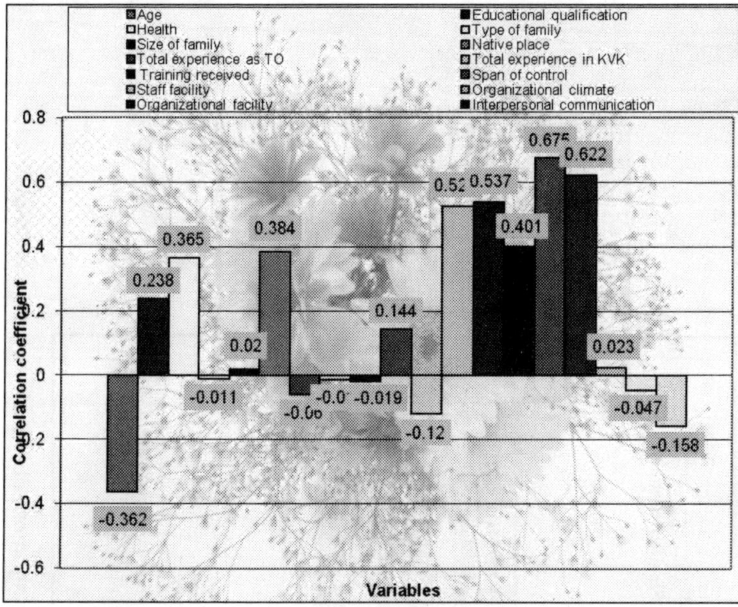

Figure 28: Relationship between the characteristics of the respondents and their extension management ability

supervising, controlling and decision making than old aged programme coordinators. The young programme coordinators due to their high level of enthusiasm to prove themselves as an important leader of organization might have taken keen interest in performing diversified activities of KVKs than old aged programme coordinators. This might be a probable reason for above finding. It is natural that the young persons are physically more powerful, active and able in using modern equipments and facilitates of communication like computer, internet and e-mail, prompt, conscious in using power, able to control difficult situations and interested in doing work in spite of risk involved than old aged person. This might be another reason for above findings.

From the above result, it can be concluded that age had negatively significant relationship with extension management ability of the programme coordinators.

Hence, the null hypothesis ($H0_1$) in this case of age was rejected and it was concluded that there was negative but significant

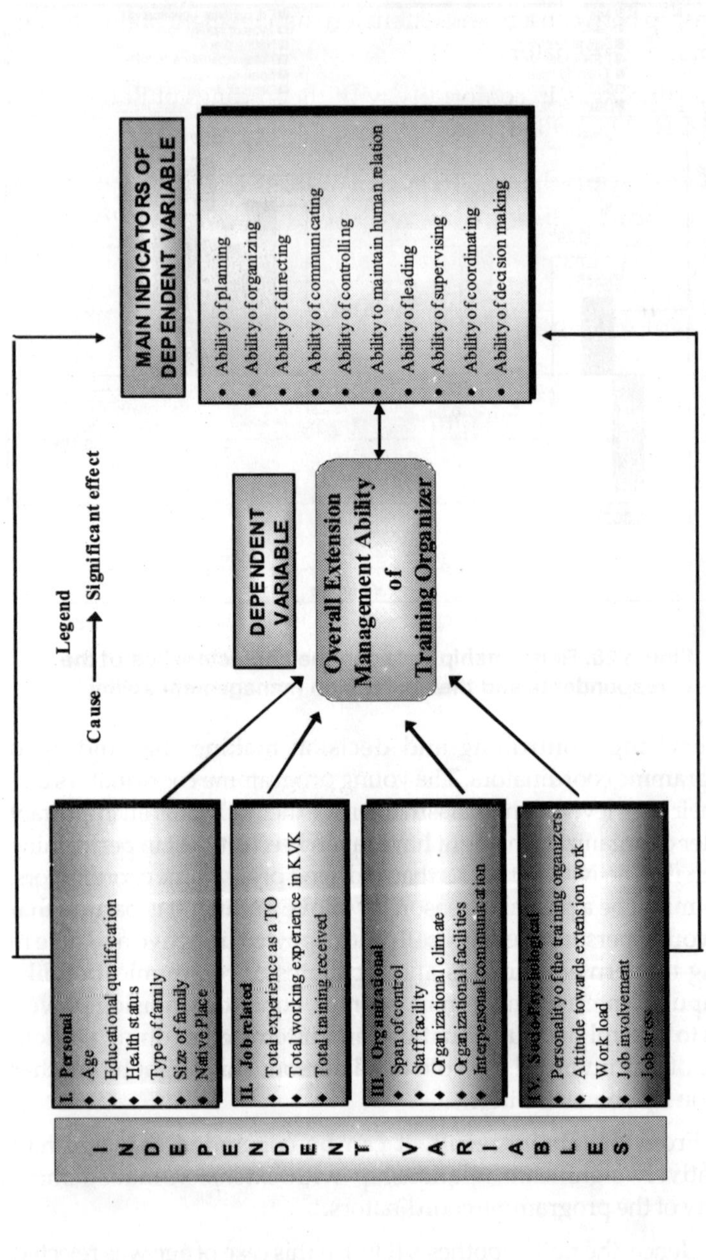

Figure 29: Conceptual model of overall extension management ability of programme Co-ordinator

relationship between age and extension management ability of the programme coordinators.

This finding is in conformity with the findings of Rani (1985) Padheria (1989) and Patel (2001).

5.3.2 Educational Qualification and Extension Management Ability

A perusal of data presented in Table 31 indicates that extension management ability of respondents had positive and significant relationship with their educational qualification. It means highly qualified programme coordinators seemed to be better than the less qualified ones. The reason for this might be that highly educated programme coordinators might have greater reception power which resulting in to more knowledge acquisition. In addition to this higher educated programme coordinators might have utilized their modern knowledge received, while earning high level of degrees to acquire ability to do better management. Further, higher education opens the world of thoughts leading towards positive thinking, better planning and proper execution. Thus, the programme coordinators with higher education might have higher management ability to work as efficient extension workers.

Hence, the null hypothesis ($H0_1$) in this case of education was rejected and it was concluded that there was positive and significant relationship between education and extension management ability of the programme coordinators.

This finding is further strengthened by the results reported by Sharma *et al.* (1988), Padheria (1989) and Patel (2001).

5.3.3 Health and Extension Management Ability

The data in Table 31 indicate that there was positive significant relationship between health of the programme coordinators and their extension management ability. It means extension management ability was observed better among those programme coordinators, who had vigorous status of health. It is natural that person having strong mental and physical status will be very positive and capable to carry out any kinds of mental, psychological, intellectual, rational and physical works. The management of any organization always needs strong will power, positive attitude, capacity to deal with any situation mentally and physically, among those who are directly involved in it. This is possible if person bears strong health. The strong health provides all kind of conducive, favorable, encouraging, helpful and contributing situations to the person for above said pre-requisites to do effective management. This might be the probable reason to have high level of management ability among those programme coordinators who had sturdy health.

From the above discussion, null hypothesis ($H0_1$) in this case of health was rejected and it can be concluded that there was significant relationship between health of programme coordinators and their extension management ability.

5.3.4 Type of Family and Extension Management Ability

The 'r' value (-0.011) seen in the Table indicates negative and non-significant relationship between the type of family of programme coordinators and extension management ability of programme coordinators. The negative relationship indicates that programme coordinators with nuclear family had high level of extension management ability than those, who were from joint family. It is understandable that the programme coordinators from nuclear family have less number of family members to take care of. This situation might have given more chance to involve and think more for office work than those who were from joint family system. In fact this trend was not up to the level of significance thus, it can be said that family type had least role on the extension management ability of the programme coordinators.

Thus, the null hypothesis ($H0_1$) was accepted and it was concluded that there was negative and non-significant relationship between type of family and extension management ability of programme coordinators.

5.3.5 Size of Family and Extension Management Ability

The number of members in the family of the programme coordinators did not play a vital role in improving extension management ability of programme coordinators. Thus, a non-significant relationship was inferred between the size of family and extension management ability of programme coordinators. It was seen that as the number of family member increased, extension management ability was decreased and vice versa but this tendency was non-significant.

Thus, the null hypothesis ($H0_1$) in case of family member of the programme coordinators was accepted and concluded that there was non-significant relationship between size of family and extension management ability.

This finding is in conformity with the finding of Patel (1993).

5.3.6 Native Place and Extension Management Ability

The relationship between native place of the programme coordinators and their level of extension management ability was positive and significant. As the method adopted for scoring this variable says that programme coordinator from urban background was given one score while programme coordinator from rural background was given two score.

The result clear-cut indicates that programme coordinators with rural native background were observed more active and commanding in managing extension activities of KVK.

As we know, majority of the Indian farmers live in rural area, thus person working as programme coordinators of KVK has to work with rural people. It is logical that the person who has rural background will have more knowledge about mentality, attitude, way of thinking, mindset, local language, customs, traditions, convention, habit, routine, tendency and problems of farming community. In addition to this, person with rural area knows and understands better about farming situation prevailing in rural area. Thus, person with such rural background can work effectively with rural people. Such kind of possibility is comparatively less in case of person who is working in rural area alongwith his urban background. This might be the probable reason to have high level of extension management ability among those programme coordinators who were from rural background.

Thus, the null hypothesis ($H0_1$) was rejected and it was concluded that there was positive and significant relationship between native place and extension management ability of programme coordinators.

5.3.7 Total Experience as to and Extension Management Ability

The correlation coefficient between total experience of programme coordinators and their level of extension management ability was -0.060 indicating non-significant relationship between them. The negative value indicates that level of extension management ability of those programme coordinators was better who had less experience of service than those of more experienced programme coordinators. It can be said that programme coordinators who were less experienced were younger than experienced one. As discussed earlier that young persons due to their elevated enthusiasm, they give the impression of being active in understanding modern way of working and system. This might be the reason to have negative relationship between these two variables.

However, 'r' value was statistically non-significant hence, the null hypothesis, ($H0_1$) was accepted and it can be concluded that total experience had non-significant relationship with extension management ability of the programme coordinators.

This finding is further strengthened by the results reported by Rani (1985), Sakaria (1988), Patel (1996) and Patel (2001).

5.3.8 Total Experience in KVK and Extension Management Ability

The correlation coefficient between total experience of programme coordinators working in KVK and their level of extension management ability was indicating non-significant relationship. The negative value indicates that level of extension management ability of those programme coordinators was observed better who had less experience of service in KVK than those of more experienced programme coordinators or it can be said that programme coordinators who were less experienced and newer than experienced one were better as extension managers. In other words, it can be said that new programme coordinators of KVK due to their prominent eagerness to be an important person of KVK gave the impression of being vigorous in proving themselves as active extension managers. This might be the reason to have negative relationship between these two variables. Nevertheless, this trend was not seen up to the considerable level.

Therefore, the null hypothesis ($H0_1$) was accepted and it can be concluded that total experience of programme coordinators with KVK had non-significant relationship with their extension management ability.

This finding is in conformity with those findings reported by Patel (1998) and Patel (2001).

5.3.9 Training Received and Extension Management Ability

The data in Table point out that there was negative insignificant relationship between training received by programme coordinators and their extension management ability. In other words it can be said that training received by the programme coordinators did not exhibit any impact in improving extension management ability of the programme coordinators. The training is said training if it shows any impact on the performance of any individual on his job. There is no meaning of any training, if it is not mental or physical skill improvement oriented. It is seen that many programme coordinators are failed to prove their ability in bringing desirable improvement on the performance of the trainees on their job. In such a situation, in spite of exposure of person under many training, one cannot expect better results. This was observed in the present study. This may be due to the reason that the general training does not support much to management aspects. Thus, the programme coordinators with general training mostly on subject matters could not improve much their skills of managerial ability of extension work.

Hence, the null hypothesis ($H0_1$) in case of training received was accepted and it was concluded that there was non-significant relationship between training received by the programme coordinators and their extension management ability.

Similar findings were observed by Rani (1985), Sakaria (1988), Patel *et al* (1998) and Patel (2001)

5.3.10 Span of Control and Extension Management Ability

The information available in Table designates that span of control has non-significant relationship with extension management ability of programme coordinators. Though the span of control has not played any significant role in increasing extension management ability, positive relationship indicates that extension management ability of those programme coordinators was better who had more workers to look after. In other words, it can be said that programme coordinators with more members to control directly by them were more active in managing KVKs. It also indicates that programme coordinator who had adopted strategy to maintain direct control and had more observation on all members of their KVKs had better extension management ability. However, this inclination was not up to the level of importance.

Thus, the null hypothesis ($H0_1$) in case of span of control and extension management ability was accepted and it can be concluded that span of control had non-significant relationship with extension management ability of programme coordinators. Similar finding was observed by Patel (2001).

5.3.11 Staff Facility and Extension Management Ability

With respect to staff facility, the data presented in Table 31 indicate that the correlation coefficient was -0.120 indicating negative non-significant relationship with extension management ability of programme coordinators. It was striking to note that the negative relationship indicates that programme coordinators with less number of staff members in their KVK proved more efficient as manager of KVK. In the present study overall management ability was measured by considering personal ability of planning, organizing, directing, communicating, controlling, leading, supervising and decision making of programme coordinators. It is natural that when you have small staff to administer, it becomes easier for you to systematize, answer, communicate, converse, correspond, direct, guide, oversee and motivate each member of the organization to perform their work effectively. This might be the probable reason to have high level of extension management ability among those programme coordinators who had less staff members in their KVKs. However, calculated 'r' value indicates that, this inclination was not observed up to the level of significance.

Hence, null hypothesis ($H0_1$) in case of staff facility was accepted and it can be inferred that there was non-significant relationship between staff facility and extension management ability of the programme coordinators.

Similar findings were reported by Patel *et al.* (1997) and Patel (2001).

5.3.12 Organizational Climate and Extension Management Ability

Table 31 shows that the correlation coefficient between organizational climate and extension management ability of programme coordinators was 0.524, indicating positive significant relationship between them. It shows that extension management ability of those programme coordinators was observed better, who had better and favourable organizational climate. The result indicates that programme coordinators with positive state of their employees' behaviours such as communication, problem solving, accountability, quality and performance, proved as favourable situation to perform as effective extension managers. In other words, programme coordinators who tried to create right environment for co-workers and simplicity by understating the true value of co-workers for the development of organization with the help of establishing proper communication, solving barriers and liability had high level of extension management ability. This might be because of simple reason that manager of any organizations can be successful if high quality work is performed by employees. Another reason for significant relationship may be that friendly, peaceful, cooperatively organizational climate might have led programme coordinators towards high level of extension management ability.

Hence, the null hypothesis, ($H0_1$) was rejected and it can be concluded that organizational climate had significant relationship with extension management ability of the programme coordinators.

This finding is in line with the findings of Rani (1985), Padheria (1989) and Patel (2001).

5.3.13 Organizational Facility and Extension Management Ability

With respect to organizational facility, the data presented in Table 31 indicate that the correlation coefficient was 0.537 indicating significant relationship with extension management ability of programme coordinators. It means that programme coordinators with good organizational facilities were proved as better managers than those of with less organizational facility.

To accomplish mandates of KVK, programme coordinators have to do many activities like training, publication of farm literature, edification and organization of demonstrations, e-extension, field visits etc. To carry out all these activities, they need many infrastructural facilities like land, laboratory, vehicles, audio-visual aids, reading materials, furniture, stationary and other communication facilities like telephone, xerox machine, fax, computer and internet connections. Thus, it was obvious that programme coordinators with all the required organizational facilities could manage their duty effectively. This might be the reason for the above relationship.

Hence, the null hypothesis, ($H0_1$) was rejected and it can be inferred that there was significant relationship between organizational facility and extension management ability of programme coordinators.

5.3.14 Interpersonal Communication and Extension Management Ability

The data in Table 31 stated that correlation coefficient of interpersonal communication with extension management ability of the programme coordinators was 0.401 indicating highly significant relationship at 0.01 level of significance. The result states that management ability of those programme coordinators was observed better who had better interpersonal communication quality.

The interpersonal communication was measured in terms of degree of contacts of programme coordinators with their heads, directors and officers of state agricultural university, state agricultural department and other officials. Thus, it is understood that programme coordinator with frequent contacts with other officers gets higher opportunity to discuss new problems, developmental issues, budge related matter, innovations and many other useful matters.

The persons working in any organization as a manager need to engage in interpersonal communication to gain knowledge about another individual. Social Penetration Theory says that we attempt to gain information about others so that we can interact with them more effectively. We can better predict how they think, feel and act if we know who they are. Programme coordinators through frequent contacts with others might have gained this ability to achieve high level of extension management ability.

Hence, the null hypothesis ($H0_1$) was rejected and it can be concluded that interpersonal communication had significant relationship with extension management ability of programme coordinators.

5.3.15 Personality and Extension Management Ability

The result shown in Table 31 indicates that the personality of the programme coordinators was significantly correlated with their extension management ability. In other words it can be said that extension management ability of programme coordinators was better who had extrovert type of personality. It is well known fact that the person who has to work to inspire several people towards constructive track should have established personality, which is not subjugated by immature pleasure force. He should illustrate maturity to deal people by understanding norms and limitations of any society. The programme coordinators, working at KVKs with definite quality to work effectively with rural people and group along with motivation, self-confidence, friendly attitude and likeness to lead social gathering were seemed to be better in their extension management ability.

Extrovert individuality helps an individual to do extension work very successfully because of permutation of useful characters and positive approach to deal with external reality and environment rather than inner feelings. In addition to this extrovert individuals are highly sociable, active, assertive, energetic, enthusiastic, outgoing, impulsive and emotionally expressive, friendly and talkative. Such positivism might have played important role to have better extension management ability among those programme coordinators who had extrovert type of personality.

Hence, null hypothesis ($H0_1$) in the case of personality of the programme coordinators was rejected and it was concluded that the personality of the programme coordinators was significantly correlated with their extension management ability.

5.3.16 Attitudes Towards Extension Work and Extension Management Ability

It could be seen from Table 21 that attitude towards extension work had positive and significant relationship with extension management ability of programme coordinators.

In any organization, the attitudes of it's employee towards their work has a great bearing on its success or failure. Thus, the favorable attitude of programme coordinators towards extension work might have psychologically favored to do better in their job work resulting in better managerial ability of extension work. This finding is supported by Rani (1985) and Patel (2001).

Thus, the null hypothesis ($H0_1$) in case of attitude of programme coordinators towards extension work was rejected and concluded that there was positive significant relationship between attitude towards extension work and extension management ability of programme coordinators.

5.3.17 Workload Perceived and Extension Management Ability

The data regarding workload perceived in Table indicate that the correlation coefficient value was 0.023 indicating non-significant with extension management ability of programme coordinators. However, positive relationship indicates that extension management ability of those programme coordinators was better who had high or reasonable workload. In spite of this, the null hypothesis, ($H0_1$) was accepted because such trend was not observed significant. Based on result it can be concluded that there was no any relationship between extension management ability of the programme coordinators and their perceived workload.

The probable reason for above findings might be that all-most programme coordinators have to perform similar kind of workload because mandates given by authority to all KVKs are more or less same.

5.3.18 Job Involvement and Extension Management Ability

In respect to job involvement, the data in Table indicate that the correlation coefficient was -0.047 indicating non-significant relationship with extension management ability. In other declaration, it can be said that job involvement did not play any role on the extension management ability of programme coordinators.

The probable reason for above findings might be same as explained in previous table that almost programme coordinators have to involve with various kind of activity to perform similarly because directives given by authority to all KVKs are relatively same.

Hence, the null hypothesis, ($H0_1$) was accepted and it can be concluded that job involvement was non-significantly correlated with extension management ability of the programme coordinators.

Similar results were reported by Veerbhadraiah (1983), Reddy and Jayaramaiah (1988) and Patel (2001).

5.3.19 Job Stress and Extension Management Ability

The data in Table 31 with respect to job stress show that, job stress was indicating negative significant relationship with extension management ability of the programme coordinators.

The job stress has been worked out as feedback of programme coordinators with regards to their emotional and/or physical pressure for their job. A hectic situation occurs when equilibrium is not experiential between demand imposed upon them by the situation and their capacity to cope with such requirements. The person who is head of KVK should have physically and mentally powerful outlook to do work in any adverse situation. He should be sufficient strong, punctual, nervousness-free, matured, accommodating and talented to do any type of job successfully with whatsoever accessible staff he has. It is difficult to do such work. Thus it was seen that extension management ability of those programme coordinators was seen better who had less job stress. Understanding this fact, it is advisable that programme coordinators working at KVKs should be given specific training to deal with high job stress.

Hence, the null hypothesis, ($H0_1$) was rejected and it can be inferred that job stress had negative significant relationship with extension management ability of programme coordinators.

This finding is in conformity with the findings of Talukdar (1984), Murthy and Somasundaram (1989) Patel *et al.* (1998) and Patel (2001).

5.4 The Functional Relationship between Extension Management Ability and Independent Variables

In the previous sub section, the relationship between independent and dependent variables were ascertained by computing correlation coefficients (r). The correlation coefficient value gives only the strength and direction of relationship between the two characters or variables, but it does not reflect on predictive ability of independent variables to the dependent variable. Hence, in order to access the amount of contribution or say influence or predictive abilities of each independent variables to the dependent variable, the stepwise regression analysis was carried out with the help of computer.

The stepwise regression as stated by Efroymson's (1962) that it is one such method, which has been widely adopted in multiple regression analysis. It has got the added advantage that at each stage of analysis; every variable is subjected to an examination for its predictive value.

The multiple correlations co-efficient (R) represents the correlation between the dependent variable and set of independent variables fitted in multiple regression equation. The co-efficient of determination (R^2) gives amount of variation accounted in dependent variables, when all independent variables were taken together in the equation and was tested with 'F' test for its significance.

The partial regression co-efficient ($by_1 j$) represents the change in dependent variable (y) with a unit change in independent variable (X_1) keeping other variables constant and it was tested with Student's 't' test for its significance.

The various independent variables had their own unit of measurement, which did not permit a comparison of the partial regression co-efficient (byi.j) values. To facilitate comparison, the partial (byi.j) values were converted into standardized partial (b'yi.j) values, which were free from the units of measurements.

The independent variables were than ranked on the basis of standard partial regression coefficient (b'yi.j) values (ignoring sign) to find out their relative importance in predicting the dependent variable.

The content of the Table 32 revealed that the variables were introduced stepwise in succession depending upon the contribution of each of the independent variable in explaining the variation on the dependent variable.

From the Table 32, it can be observed that out of 19 independent variables only 3 were acquainting influence on the extension management ability of the programme coordinators. All the three

Table 32: Step-wise multiple regression analysis of extension management ability of programme coordinators

Sl.No.	*Independent Variable*	*Multiple Co-relation Co-efficient (R)*	*Co-efficient of Determination (R^2)*	*'F' Values*	*Partial regression Co-efficient (b)*	*'t' Value*	*Standard Partial Regression Co-efficient (SPRC)*	*Rank*
1.	Attitude towards extension work	0.675	0.455 (45.50 per cent)	131.952 **	1.700	11.487 **	0.443	I
2.	Personality of the programme coordinators	0.736	0.542 (54.20 per cent)	92.781 **	0.383	5.447 **	0.285	II
3.	Organizational climate	0.746	0.557 (55.70 per cent)	65.376 **	0.236	2.320 *	0.155	III

* Significant at 0.05 level of probability.

** Highly significant at 0.01 level of probability.

variables together were contributing 55.70 per cent variation as indicated by 'R^2' value for the extension management ability.

It can be inferred that 45.50 per cent variation in extension management ability of programme coordinators of KVKs was contributed by attitude toward extension work. However, attitude towards extension work + personality of the programme coordinators accounted for 54.20 per cent, where as, attitude towards extension work + personality of the programme coordinators + organizational climate accounted 55.70 per cent variation in extension management ability of programme coordinators of KVKs. The R^2 values at each stage of step up regression were found to be significant at 0.01 level of probability.

The partial 'b' values of these three variables were converted into standard partial 'b' values, which were 0.443 for attitude towards extension work, 0.285 for personality of the programme coordinators and 0.155 for organizational climate. The 't' value or partial 'b' were significant in case of all the three independent variables.

According to the highest to lowest standard partial 'b' the rank order were given first attitude towards extension work, second to personality of the programme coordinators and third to organizational climate.

Hence, null hypothesis ($H0_2$), in case of independent variables like attitude towards extension work, personality of programme coordinators and organizational climate was rejected and for remaining independent variables, it was accepted.

It can be inferred from the above results that independent variables such as attitude towards extension work, personality of the programme coordinators and organizational climate contributed 55.70 per cent variation in extension management ability of the programme coordinators.

The result shown in above Table indicates that the extrovert personality of the programme coordinators with their positive attitude towards extension activity and favourable organizational climate together played major role in making programme coordinators efficient manager of the training centres. It is natural that man with such positivism will always get chance to use his assertive and mature personality with positive approach to deal with internal and external reality and environment through positive inner feelings. It is also obvious that programme coordinators with favourable state of mind within themselves and positive organizational climate will get enough confidence to work as an efficient manager in any organizations. This might be the probable reason to receive major contribution accounted by the variables like attitude towards extension work, personality and organizational climate on the extension management ability of programme coordinators of KVKs.

The findings are suggestive of the fact that for increasing the extension management ability of the programme coordinators, proper training to build up positive feelings towards extension activity and extrovert personality should be organized for programme coordinators, at the same time he and his staff members should also be trained in such a way that conducive organizational climate can be maintained to get best results.

5.5 Constraints Faced by Programme Coordinators in Managing KVK and Suggestions Offered by Programme Coordinators to Overcome the Constraints

5.5.1 Constraints Faced by Programme Coordinators in Managing KVK

In the present study, some constrains faced by the programme coordinators in managing KVK were also studied. They were requested to express the constraints faced by them. These constraints were ranked according to their importance based on mean score and are presented in Table 33.

A glance of the Table 33 shows that no promotion channel has been arranged for the staff of KVK, many post of the subordinate staff lying vacant, intermittent flow of the funds from organization to KVK head for carrying out mandated activities, fund for the extension activities and FLDs is not released in time by the council, frequent transfer of the staff from KVK to other department are the major constraints faced by the programme coordinators in managing the KVK. Further they opined that most of the staff is engaged with organizational work or say other than KVK activities, inadequate infrastructure facilities such as building, fencing, tube well, vehicle etc., unnecessary involvement of organizational head in taking decision regarding planning, execution and implementations of the KVK activities, inadequate required educational qualification among the subordinate staff, local organizational rules do not allow the KVK staff to work freely according to the guidelines of council were the other constraints experienced and expressed by the programme coordinators to restrict their extension management ability.

Table 33: Constraints faced by programme coordinators in managing KVK

n = 160

Sl.No.	Constraints	Mean Score	Rank
1.	Many posts of the subordinate staff lying vacant	1.79	II
2.	Majority of the subordinate staff not fulfilling the required educational qualification	1.31	IX
3.	Frequent transfer of the staff from KVK to other department	1.52	V
4.	Fund for the extension activities/FLDs is not released in time by the council	1.58	IV
5.	Most of the staff remains engaged with local organizational work and less for the KVK activities	1.41	VI
6.	No promotion channel for the KVK staff	1.93	I
7.	Inadequate infrastructure facilities such as building, fencing, tube well, vehicle etc.	1.38	VII
8.	Unnecessary involvement of organizational head in taking decision regarding planning, execution and implementations of the KVK activities.	1.35	VIII
9.	Unfair interference of the local organizational head should be removed	1.29	X
10.	Intermittent flow of the funds from organization to KVK head for carrying out mandated activities	1.59	III
11.	Pressure from the local politician to fetch more benefits from KVK schemes to their own jurisdiction	1.13	XI
12.	Local organizational rules do not allow the KVK staff to work freely according to the guidelines of council	1.31	IX

Promotion encourages employees to view the organization as one offering them career growth. An incentive like promotion increases participation rate of employees and helps to create healthy completion to develop organization. It is needed to adhere healthy behaviors of employee. The purpose of the incentive is to encourage employees to adopt positive behaviors or maintain an existing positive behavior. Everyone who achieves a goal or maintains a behavior should receive something. Many organizations also provide incentives merely for active and healthy participation of employee in the development of organizations. Promotion helps in making organization more effective and enhances responsiveness of the employees. For this reason, if channel of promotion does exist in KVK setup, authority should take care of this constraint expressed by the programme coordinators.

The another important constrain faced by the programme coordinators, was issue of the vacant posts, it is natural that to perform all the activities of KVK effectively, sufficient number of subordinate staff is needed by the manager or any organizers, understanding this, urgent appointment of subordinate staff on the vacant posts was realized by the programme coordinators. The high performance team of human is one of the most important resources in development of any public dealing organizations. The organization like KVKs are involved in many activities, so to carry out work as per the mandates, programme coordinators need necessary active human resources. One should not expect best result from programme coordinators without enough and skillful human resources. In spite of positive attitude towards extension activity and mature personality, programme coordinators cannot do all kinds of work without skillful force of colleagues. The organizations that pay attention to the key factor of giving professional force of co-workers to the managers can create a positive and productive environment for their organizations, that will drive overall performance and long-term success.

The high performance teams with the quality of autonomous team- work attitude and self-directed work nature can perform their various roles effectively. Such kinds of co-workers are highly motivated and self-sufficient - drawing on their own resources and expertise. Rather than providing direction to high performance teams, managers act as consultants or resources to be utilized, as and when needed. Their autonomy that they have over their work activities enhances their ability to influence management's decisions on a variety of operational and strategic issues. High-performance teams of co-workers usually achieve a high level of motivation. Motivation, in this regard is translated into zeal, drive or enthusiasm. They are impelled, intrinsically to improve their performance. This high-level of motivation is achieved mainly because members of high-performance oriented co-workers feel that they are individually powerful, in control, highly responsible, eventually accountable, genuinely committed, highly dependable, highly empowered with the requisite degree of professional autonomy, explicitly trusted and highly respected.

Realizing this, constraints like inadequate number of subordinate staff, lack of qualified staff and unnecessary engagement of staff in other than organizational activities were faced by the programme coordinators of KVKS.

All over India, KVKs are being run by SAUs, NGOs, Vidyapiths and ICAR research stations, it is therefore, programme coordinators have to depend on the concern authority to utilize funds of KVK because funds are not being given directly to them. Within this circumstance, untimely availability and improper utility of needed funds were understood as constraints by the programme coordinators. In addition to this, engagement of KVK staff to do less work for KVK and more for local organizations, intermittent flow of the funds from organization to KVK head for carrying out mandated activities, untimely release of fund for the extension activities/FLDs, pressure from the local politician to fetch more benefits from KVK schemes to their own jurisdiction and dependency on the administrative procedure of local organizational to work freely according to the guidelines of council were another constrains faced by the programme coordinators.

Hence, null hypothesis ($H0_3$) was rejected and it can be concluded that programme coordinators faced some constraints in effective management of extension activities of KVKs.

5.5.2 Suggestions Offered by Programme Coordinators to Overcome the Constraints in Managing KVK

In the present study, after studying some constraints faced by the programme coordinators in managing KVK, they were also requested to express their valuable suggestions based on their practical experiences. These suggestions were felt important to study to provide feed back to the concerned authority for the effective results of existing KVKs as well as to provide some guideline for those KVKs which are under considerations to establish allover India. The suggestions were collected in three continuums and their mean scores were worked out to understand its importance. The suggestions were also ranked according to their mean score and are presented in Table 34.

Table 34: Suggestions offered by programme coordinators to overcome the constraints in managing KVK

n = 160

Sl.No.	*Suggestions*	*Mean Score*	*Rank*
1.	Vacant posts especially technical including programme coordinators should be filled up immediately	1.63	I
2.	Disciplines as well as qualification should be considered while appointing technical staff	1.52	II
3.	Career advancement scheme for the KVK staff should be implemented and frequent transfer of the KVK technical staff should be avoided	1.40	III
4.	Demonstration unit/training laboratory as well as infrastructural premises should be well equipped	1.36	IV
5.	KVK grant should be directly released to the KVK account for timely operation. Additional grant should be provided to the KVK for carryout extension activities	1.32	V
6.	Sufficient grant especially for extension activities should be provided and such grant should be released timely	1.30	VI
7.	Power and authority should be delegated to the programme coordinators adequately, considering their responsibility	1.27	VII
8.	Vehicle facility especially for carrying extension activities in village situation is highly essential	1.24	VIII
9.	The KVKs should be restricted to perform only extension tasks. Administration and extension work should be separated accordingly	1.21	IX
10.	Modern information and communication technology should be made available to all KVKs	1.19	X
11.	A strong coordination and linkage on permanent basis should be established between research institutions and KVKs	1.17	XI

Contd...

Table 34–Contd...

Sl.No.	Suggestions	Mean Score	Rank
12.	All the necessary infrastructural facility and educational facility for the staff should be made easily available	1.14	XII
13.	There should be a provision for work motivation and incentives for the staff of KVKs	1.11	XIII
14.	Frequent transfer of the staff should be restricted.	1.09	IXV
15.	Mechanism of strong linkages and coordination with other extension agencies is necessary	1.07	XV
16.	An adequate personnel management policy should be adopted	0.62	XVI
17.	Standard and uniform evaluation system for extension work should be suggested by the ICAR to all the KVKs	0.38	XVII

Table 34 reveals major suggestions given by programme coordinators in performing their work as extension mangers of KVKs. These suggestions are listed as bellow.

1. Vacant posts especially technical including programme coordinators should be filled up immediately. Without adequate technical staff, one cannot achieve the objectives of the KVK successfully.
2. Disciplines of subject as well as needed qualification should be maintained, while appointing technical staff. Without appropriate discipline/qualification/specialization, quality of extension work may suffer a lot.
3. Career advancement scheme for the KVK staff should be implemented and frequent transfer of the KVK technical staff should be avoided.
4. Demonstration-unit/training-laboratory as well as infrastructural premises should be well equipped. Without the proper practical training facilities, one cannot do practice skillfully.
5. KVK grant should be directly released to the KVK for timely operation. Additional grant should be provided to the KVK to carryout extension activities *e.g.* Agricultural Fairs, Exhibition, celebration of different days like farmers' day,

world food day etc. and maintenances of infrastructural facility.

6. Sufficient grant especially for extension activities should be provided and such grant should be released timely. Insufficient grant and delay in release of grant largely influence on planning and implementing the extension programme timely.
7. Power and authority should be delegated to the programme coordinators adequately, considering their responsibility. Decentralization of power and authority may be helpful to maintain the standard/quality or work and to carry out the responsibilities effectively.
8. Vehicle facility especially for carrying extension activities in village situation is an essential requirement, because extension work of KVKs is entirely rural based.
9. The KVKs should be restricted to perform only extension tasks, administration and extension work should be separated accordingly.
10. Modern information and communication technology such a s computer, internet, fax, zerox, mobile phone, etc. should be made available to all KVKs, which are very essential for the development of extension management.
11. A strong coordination and linkage on permanent basis should be established between research institutions and KVKs with the help of the modern information/ communication technology which will reduce the communication gap between them.
12. All the necessary infrastructural facility and educational facility for the staff shouid be made easily available which indirectly influences in a positive manner on extension management work of KVKs.
13. There should be a provision for work motivation and incentives for the staff of KVKs, which may result in effective and efficient management of extension work.
14. Frequent transfer of the staff is harmful. It adversely affects the planning and implementation of pre-decided extension programmes. Thus, model criteria for transfer

of the staff should be designed by the ICAR especially for the KVKs working under SAU and ICAR.

15. Strong linkages and coordination with other extension agencies is necessary for effective management of extension programme.
16. An adequate personnel management policy should be adopted for better management of extension work of KVKs.
17. Standard and uniform evaluation system for extension work should be suggested by the ICAR to all the KVKs.

Table 34 reveals that major suggestions given by programme coordinators in performing their work as extension mangers of KVKs were related to appointment of staff, sufficient and timely major and additional grants, infrastructural facility, incentives, vehicle facility, career advancement, restriction on frequent transfer, modern information communication technology, linkage with other organizations and personnel management policy. To get best results through KVKs, their practical suggestions should be considered by the concerned authority. In addition to this, while establishing new KVKs allover India, programme coordinators of new KVKs should be provided all the facilities to run the activities of KVK as per the desired goals and mandates.

Chapter 6

Summary and Conclusion

This chapter includes in a nutshell the description of summary, conclusion, problems and suggestions for future thrust.

6.1 Summary

The vigorous attempt was realized by the Education Commission (1964-66) to establish a specialized institution to provide vocational education in agriculture and allied fields at the pre and post matriculation levels to cater to the training needs of a large number of boys and girls coming from rural areas. The recommendation of Education Commission was thoroughly discussed during 1966-1972 associating the Ministry of Agriculture, Ministry of Education, Planning Commission, ICAR and other institutions. Finally, the ICAR mooted the idea of establishing Krishi Vigyan Kendra (Farm Science Centre) as innovative institutions for imparting vocational training to the farmers and field level extension functionaries. As a result of this, in 1973 a committee headed by Dr. Mohan Singh Mehta was constituted and based on its report in 1974, it was decided to have Krishi Vigyan Kendras as Farm Science Centres for speedy transfer of technology to the farmer's fields. At present, about 504 KVKs working in India have been designed to have vocational training for the practicing farmers in service field to those who seek self-employment. The KVK as a grass root level institution has to meet this gap, which is so vital for technical literacy and rapid transfer of technology.

Krishi Vigyan Kendra is such an organization, which deals with particular social system within which a number of groups of people like; specialists, administrators, technicians, farmers, extension workers, labourers and other beneficiaries are working and involving. Each group has its own characteristics, attitudes, traditions, values and forms of behaviour. The people for whose benefits, the extension activities are designed may not be readily interested due to multiplicity of factors like; likeness, inhibitions, taboos, suspiciousness, reservations and prejudices. It is therefore, expected that manager of the KVK should have enough understanding of such social aspects to get best results out of KVKs. Thus, effectiveness of the KVK mainly depends on how it is efficiently and successfully managed by its heads or programme coordinator. The programme coordinators have to perform multifold roles of planner, organizer, director, communicator, controller, human relation maintainer, supervisor, coordinator, leader and decision-maker. It was, therefore, felt necessary to examine existing extension management abilities of programme coordinators of the KVKs. For this, present study on extension management abilities of programme coordinators of KVK was undertaken with following objectives.

1. To study personal, job related, organizational and socio-psychological characteristics of the programme coordinators of KVKs.
2. To measure the managerial ability of programme coordinators of KVKs.
3. To find out the relationship between selected independent variables and managerial ability of the programme coordinators of KVKs.
4. To estimate the functional relationship between managerial ability and various independent variables.
5. To study the constraints faced by programme coordinators of KVKs in the management of different activities of KVKs and seek their suggestions.

6.2 Review of Literature

On the basis of literature reviewed related to the study, a theoretical orientation was developed and various concepts were operationalized. Based on assumption, tentative conceptual model was developed and with the help of theoretical orientation, three statistical null hypotheses were also formulated in view of the above objectives.

6.3 Methodology

The present study was carried out in all eight zones of India, where Krishi Vigyan Kendras are in operation in transfer of technology process. The Ex-post facto research design was applied for the study. A random sample of total 160 programme coordinators of KVKs from the different zones of India was selected for this study. The methodological procedure consisted of dependent and independent variables. The independent variables studied were Age, Educational Qualification, Status of Health of programme coordinator, Type of Family, Size of Family, Native Place, Experience as a programme coordinator of KVKs, Total working experience in KVK, Total Training received, Span of Control, Availability of Technical Staff, Organizational climate, Organizational facilities, Inter-Personal communication, Personality of the Programme coordinators, Attitude towards extension work, Work load, Job involvement and Job stress. These variables were measured by using well-developed scales developed by various social scientists with due modifications. The extension management ability of the programme coordinators of KVK was dependent variable, which was measured by using the scale developed by Patel (2001) with slight modifications. The data were collected by personal interview as well as mail questionnaire. The data thus, collected were classified, tabulated and analyzed in order to make the finding meaningful. The statistical measures, such as percentage, mean score, standard deviation, coefficient of correlation, stepwise regression and standard partial regression coefficient were used.

6.4 Hypotheses of the Study

$H0_1$: There will not be any relationship between personal, job oriented, socio-psychological and organizational character of programme coordinators of KVKs and their extension management ability.

$H0_2$: There will not be functional relationship between extension management ability of programme coordinators and various independent variables.

$H0_3$: There will not be any constraints faced by programme coordinators of KVKs in managing KVK activities.

6.5 Major Findings

The following conclusions were drawn based on findings of the study:

6.5.1 Profile of the Programme Coordinators of KVK

1. Majority (58.75 per cent) of the programme coordinators had old age, followed by 31.88 per cent with middle age, while only 9.37 per cent of them had young age.
2. Majority (71.88 per cent) of the programme coordinators were educated up to doctorate level, followed by 19.37 per cent with master degree, while few of them (8.75 per cent) were educated up to bachelor degree only.
3. Two fifth (40.00 per cent) of the programme coordinators realized that their health was good, followed by 36.25 per cent and 21.25 per cent of the programme coordinators perceived that they were having average health and very good health, respectively. Equal percentages (1.25 per cent) of the programmé coordinators felt that their health was poor or very poor.
4. Majority (80.00 per cent) of the programme coordinators had the nuclear type of family, while only 20.00 per cent of them had joint type of family.
5. Majority (60.00 per cent) of the programme coordinators had up to four members in their family, while 32.00 per cent of them had five to eight members in their family and only 8.00 per cent had above eight members in their family.
6. Majority (67.5 per cent) of the programme coordinators of KVKs had urban background, whereas 32.5 per cent of the programme coordinators of KVKs were from the rural area.
7. Majority (73.76 per cent) of the programme coordinators had medium level of total experience. The low and wide intensity of experience was observed among 14.37 per cent and 11.87 per cent of the programme coordinators, respectively.
8. More than half (51.88 per cent) of the programme coordinators had medium level of total experience in KVK on a different positions, whereas 25.00 per cent and 23.12 per cent of them had wide and low level experience, respectively.
9. Three-fourth (76.25 per cent) of the programme coordinators have received less training, whereas slightly less than one fourth (23.75 per cent) of them received high training.

10. Majority (80.62 per cent) of the programme coordinators had medium level of span of control, whereas 13.75 per cent and 5.63 per cent of them had low and high level of span of control, respectively.
11. Exactly half (50.00 per cent) of the programme coordinators opined that the strength of staff provided them was medium, whereas, 26.25 per cent of programme coordinators opined that existing staff strength was as per the requirement, whereas 23.75 per cent of them felt that insufficient staff was provided to their KVKs.
12. Majority (69.38 per cent) of the programme coordinators had perceived medium level of favourable organizational climate, while 16.25 per cent and 14.37 per cent of them had low and high level of favourable perception for their organizational climate.
13. Majority (74.38 per cent) of the programme coordinators felt that existing organizational facility with in their KVK was medium. On the other hand,low and high levels of organizational facility were perceived by 5.00 per cent and 20.62 per cent of the respondents, respectively.
14. Majority (58.13 per cent) of the programme coordinators had medium level interpersonal communication, whereas, 23.75 per cent of them had low level and only 18.12per cent of them had high level of interpersonal communication.
15. Majority (76.88 per cent) of the programme coordinators exhibited extrovert personality.
16. Majority (78.75 per cent) of the programme coordinators of KVKs had favorable to highly favorable attitudes towards extension work.
17. Majority (60.00 per cent) of the programme coordinators had perceived desirable workload, whereas, 39.38 per cent of them perceived heavy workload, only 00.62 per cent of them had perception of light workload.
18. Majority (80.00 per cent) of the programme coordinators were reasonably favourable, followed by 11.25 per cent with less and 8.75 per cent with high favourable towards their job.

19. Majority (73.12 per cent) of the programme coordinators experienced medium level of job stress, whereas 13.75 per cent with low level of job stress and 13.13 per cent of them had perception of high level of job stress.

6.5.2 Extension Management Ability of the Programme Coordinators

1. Slightly more than half (51.25 per cent) of the programme coordinators had high level of planning ability, followed by 33.75 per cent and 15.00 per cent with medium and low level of planning ability, respectively.
2. Majority (85.00 per cent) of the programme coordinators had medium to high level of organizing capacity in relation to extension management ability.
3. Majority of the (76.87 per cent) programme coordinators had high level of ability of directing their subordinates, followed by 13.75 and 9.38 per cent of them were with medium and low level of ability of directing their subordinates for efficient extension management activities.
4. Majority of the programme coordinator (58.13 per cent) had high level of ability of communicating, followed by 21.25 per cent of them with low and 20.62 per cent of them were with medium level of ability of communicating, respectively.
5. Slightly less than half (48.12 per cent) of the programme coordinators had high level of controlling power, followed by 28.12 per cent and 23.76 per cent of them were with low and medium level of controlling power, respectively.
6. Great majority (91.87per cent) of the programme coordinators had medium to high level of capacity to maintain human relation, followed by only 8.13 per cent of them had low level of human relation maintaining capacity.
7. Majority (61.88 per cent) of the programme coordinators had high level of feeling of lead taking behaviour, followed by 20.00 per cent and 18.12 per cent of them were with medium and low level of feeling of lead taking behaviour, respectively.

8. Majority (71.26 per cent) of the programme coordinators indicated high level of supervising capacities, followed by 20.62 per cent and 8.12 per cent of them had medium and low level of supervising capacities, respectively.
9. Majority of the programme coordinators were found with high level of coordinating ability (62.50 per cent), followed by 26.87 per cent of them with low and only 10.63 per cent of them with medium level of ability of coordination in concern with personnel working in the KVKs.
10. Slightly less than half (45.00 per cent) of the programme coordinators had high level of decision making capacity, followed by 40.62 per cent of them with medium and only 14.38 per cent of them with low level of decision making capacities, respectively.
11. Majority (68.12 per cent) of the programme coordinators were observed with the medium level of overall extension management ability, while 18.75 per cent with high and 13.13 per cent with low level of overall extension management ability.

6.5.3 Relationship between the Characteristics of the Programme Coordinators and their Extension Management Ability

1. The personal variables like education, health status and native place of programme coordinators were observed positively, while age was observed negatively significant with their extension management ability. The variables viz., size of family and type of family were non-significant with extension management ability of programme coordinators.
2. The Job related variables like total experience as TO, total experience in KVK and training received were negatively non-significant with extension management ability of programme coordinators.
3. The organizational variables of programme coordinators like organizational climate, organizational facility and interpersonal communication were observed positively significant with their extension management ability.

While, other organizational variables like span of control was observed non-significant, only staff facility was observed negative but significant with their extension management ability.

4. The socio-psychological variables of programme coordinators like personality and attitudes towards extension work were observed positively significant, while job stress was observed negatively significant with their extension management ability. The other socio-psychological variables; workload and job involvement observed non-significant with extension management ability.

6.5.4 The Functional Relation between Extension Management Ability and Independent Variables of the Programme Coordinators

1. The 45.50 per cent variation in extension management ability of training organizers of KVKs was contributed by attitude toward extension work. However, attitude towards extension work + personality of the training organizers accounted for 54.20 per cent where as, attitude towards extension work + personality of the programme coordinators + organizational climate accounted 55.70 per cent variation in extension management ability of programme coordinators of KVKs.

6.5.5 The Constraints Faced by the Programme Coordinators

1. Lack of promotion channel for the staff of KVK, many post of the subordinate staff lying vacant, intermittent flow of the funds from organization to KVK head for carrying out mandated activities, pressure from the local politicians to fetch more benefits from KVK schemes to their own jurisdiction, funds for the extension activities and FLDs are not released in time by the council, frequent transfer of the staff from KVK to other department are the major constraints faced by the training organizers in managing the KVK.
2. Further they opined that most of the staff is engaged with

organizational work or say other than KVK activities, lack of adequate infrastructure facilities such as building, fencing, tube well, vehicle, unnecessary involvement of organizational head in taking decision regarding planning, execution and implementations of the KVK activities, inadequate required educational qualification among the subordinate staff, local organizational rules do not allow the KVK staff to work freely according to the guidelines of council are the other constraints experienced and expressed by the programme coordinators to restrict their extension management ability.

6.5.6 The Suggestions Expressed by the Programme Coordinators to Overcome Constraints

1. The major suggestions given by programme coordinators in performing their work effectively as extension mangers of KVKs were related to appointment of qualified technical staff, sufficient and timely availability of main and additional grants, infrastructural facility, incentives, vehicle facility, career advancement, restriction on frequent transfer, modern information communication technology, linkage with other organizations and personnel management policy. To get best results through KVKs, their practical suggestions should be considered by the concerned authority. In addition to this, while establishing new KVKs allover India, programme coordinators of new KVKs should be provided all the facilities to run the activities of KVK as per the desired goals and mandates.

6.6 Empirical Model

Base on results of relationship of independent variables with extension management ability of programme coordinators, an empirical model has been prepared and it has been given in Fig. 31.

It was seen that personal variables of programme coordinators of KVKs such as; young age, higher education, vigorous status of health and rural native place, while organizational variables like; conducive organizational climate, needed organizational facility and better interpersonal communication, whereas socio-psychological variables for instance; extrovert personality and positive attitudes towards extension work and low level of job stress played significant role on their extension management ability.

6.7 Implications

The present study was highlighted direction on the new area in which, research work was needed to carry out. The following implications emerged from this study.

1. A definite approach to craft India as developed nation in the shortest possible time is realized by a rapid agricultural growth. To accomplish the admired dream of father of nation; Mahatma Gandhi, we have to work with commitment in taking over newer information and technology to our millions of farmers who work hard vigorously in the sun to feed the nation. The world is shifting very rapidly. Technology of today becomes outdated tomorrow and agriculture is not exception to this. Agriculture plays an imperative role in Indian economy. Today in first line of transfer of technology mission, Krishi Vigyan Kendra needs to play a role of key extension system to develop the farming communities.
2. The findings of the study reveal that, to manage and run KVKs as per its mandate, there is need to appoint young, talented, well qualified, dedicated, enthusiast programme coordinators to reinforce role of KVKs.
3. The results indicates that programme coordinators with rural native background were observed more active and commanding in managing extension activities and KVK, thus it is suggested that person with rural background or feelings to work in rural and semi urban area should be appointed to work as programme coordinators of KVKs. A Programme coordinator with rural background or feeling to work with farmers knows and understands better about farming situation prevailing in rural area.
4. During the study it was seen that training received by the training organizers did not exhibit any impact in improving their extension management ability. The training is said training if it shows any impact on the performance of any individual on his job. There is no meaning of any training, if it is not mental or physical

skill improvement oriented. It is seen that many training organizations are failed to prove their ability in bringing desirable improvement on the performance of the trainees on their job. In such a situation, in spite of exposure of person under many training one cannot expect better results. This was observed in the present study. This may be due to the reason that the general training does not support much to management aspects. Thus, the training organizers with general training mostly on subject matters will not improve their skills of managerial ability of extension work. Thus, there is need to give special trainings which can prepare programme coordinators as good manager.

5. The positive relationship indicates that extension management ability of those programme coordinators was seen better who had more workers to look after. The programme coordinators with more members to control directly by them were more active in managing KVKs. It indicates that training organizer who had adopted strategy to maintain direct control and observation on all members of their KVKs had better extension management ability. Thus, it is recommended that programme coordinators should try to maintain direct contact with each member of KVK for effective management.
6. It was seen that extension management ability of those training organizers was observed better who had better and favourable organizational climate. The programme coordinators with positive state of their employees' behaviours such as communication, problem-solving, accountability, quality and performance got favourable situation to perform as effective extension managers. In other words training organizers who tried to create right environment for co-workers and simplicity by understating the true value of co-workers for the development of organization with the help of establishing proper communication, solving barriers and liability had high level of extension management ability. Thus, programme coordinators are recommended to maintain conducive organization climate to improve their level of extension management ability to manage KVKs in effective manner.

7. To accomplish mandates of KVK, programme coordinators have to perform many activities like training, publication of farm literature, edification and organization of demonstrations, e-extension, field visits etc. For this, they need many infrastructural facilities like land, laboratory, vehicles, audio-visual aids, reading materials, furniture, stationary and other communication facilities like telephone, zerox machine, fax, computer and internet connections. Thus, it is recommended to provide all the facilities to each programme coordinator to manage their duty effectively.
8. The interpersonal communication is key component in development of any organization. It is degree of contacts of head of organization with other officials. It is needed to gain knowledge from another individual to predict how they think, feel and act. Such act provides ability to achieve high level of extension management ability. Thus, it is recommended to develop better interpersonal and interorganizational communication ability among the programme coordinators of KVKs.
9. The extension management ability of programme coordinators was observed better who had extrovert type of personality. The programme coordinators, working at KVKs with definite quality to work effectively with rural people and group along with motivation, self confidence, friendly attitude and likeness to lead social gathering, were seemed to be better in their extension management ability. It is thus, dreamed that each KVK of India should have such kind of programme coordinators. For this psychological training should be given to programme coordinators to develop extrovert individuality to make them able to deal with external reality and environment rather than inner feelings. In addition to this, such trainings help them to be highly sociable, active, assertive, energetic, enthusiastic outgoing, impulsive and emotionally expressive and open-minded.
10. The attitude towards extension work was seen positive and significant with extension management ability of programme coordinators. In any organization, the

attitudes of it's employee towards their work has a great bearing on its success or failure. Thus, some strategy should be there to select and appoint person as programme coordinator of KVK with favorable attitude towards extension work.

11. The job stress was seen negative significant with extension management ability of the programme coordinators. The job stress is feedback of training organizers with regards to their emotional and/or physical pressure for their job. A hectic situation occurs when equilibrium is not experiential between demand imposed upon them by the situation and their capacity to cope with such requirements. The person who is head of KVK should have physically and mentally powerful outlook to do work in any adverse situation. He should be sufficient strong, punctual, nervousness free, matured, accommodating and talented to do any type of job successfully with whatsoever accessible staff he has. It was seen that extension management ability of those programme coordinators was seen better who had less job stress. Understanding this fact it is advisable that programme coordinators working at KVKs should be given specific training to deal with high job stress.

12. The results indicates that variables such as attitude towards extension work, personality of the programme coordinators and organizational climate contributed 55.70 per cent variation in extension management ability of the programme coordinators, thus authority of KVK should try to do something considering these variables to create ideal and conducive situation in KVKs.

13. In the present study, some constrains faced by the programme coordinators in managing KVK were also studied. It was realized that no proper promotion channel has been arranged for the staff of KVK, many post of the subordinate staff are lying vacant, intermittent flow of the funds is not that effective from organization to KVK head for carrying out mandated activities, pressure from local politicians to fetch more benefits from KVK schemes to their own jurisdiction, fund for the extension activities

and FLDs is not released in time by the council and frequent transfer of the staff from KVK to other department are the major constraints, which need to solve immediately.

14. During the study it was seen that there is perception of training organizers that most of the staff is engaged with organizational work (other than KVK activities), at the same time inadequate infrastructure facilities such as building, fencing, tube well, vehicle etc. were also realized. Such problems need to resolve.
15. The KVKs are being managed by various organizations, so unnecessary involvement of organizational heads in taking decision regarding planning, execution and implementations of the KVK activities needs be resolved.
16. The another important constrain faced by the programme coordinators, was issue of the vacant posts, It is natural that to perform all the activities of KVK effectively, sufficient number of subordinate staff is needed by the manager of any organizers, understanding this, urgent appointment of highly qualified subordinate staff on the vacant posts should be realized. The high performance team of human is one of the most important resources in development of any public dealing organizations. The organization like KVKs are involved in many activities, so to carry out work as per the mandates, programme coordinators need necessary human resources. One should not expect best result from programme coordinators without enough and skillful human resources. Inspite of positive attitude towards extension activity and mature personality, programme coordinators cannot do all kinds of work without skillful force of colleagues. Organizations that pay attention to the key factor of giving professional force of co-workers to the managers can create a positive and productive environment for their organizations that will drive overall performance and long-term success.
17. All over India, KVKs are being run by SAUs, NGOs, Vidyapiths and ICAR research stations, it is therefore, programme coordinators have to depend on the concern authority to utilize funds of KVK because funds are not being given directly to them. Within this circumstance,

untimely availability and improper utility of needed funds were understood by the programme coordinators. In addition to this, engagement of KVK staff to do less work for KVK and more for local organizations, intermittent flow of the funds from organization to KVK head for carrying out mandated activities, untimely release of fund for the extension activities/FLDs, pressure from the local politician to fetch more benefits from KVK schemes to their own jurisdiction and dependency on the administrative procedure of local organizational to work freely according to the guidelines of council were another considerable issues to resolve.

18. There should be a provision for work motivation and incentives for the staff of KVKs.
19. The KVKs should be restricted to perform only extension tasks. Administration and extension work should be separated accordingly.
20. Modern information and communication technology such as Computer with all the software and accessories, internet, fax, zerox, mobile phone, printer, etc. should be made available to all the KVKs which are very essential for the modern development of extension management. It will be useful to establish strong coordination and linkage on permanent basis between research institutions and KVKs.
21. All over India, most of the KVKs are being established in rural or semi- urban areas, in this situation, dependent family members and children of staff working in KVKs face difficulty of good quality education, entertainment and some other facilities, authority should do something for this key issue and the necessary infrastructural facility and educational facility should be made easily available.

6.8 Suggestions for Future Thrust

In the light of findings of the study, the following studies can be undertaken to explore more in the area of internet exposure.

1. The present study was conducted in 160 KVKs of India, as this study is the first of its kind to know the extension management ability of training organizers. This study

opens possibilities to carry out such studies on other programme coordinators of remaining KVKs of India.

2. The study was carried out under certain limitations of time and resources available with researcher, covering only programme coordinators of KVKs. However such study can be carried out among other staff members of KVKs of India also.
3. This investigation was limited to only extension management ability of programme coordinators, but other aspects like impact of internet exposure on the personality development of programme coordinators can also be included in new study.
4. Such study can be conducted on the extension educationists working in Agriculture, Horticulture, Animal Husbandry and other developmental departments of the Gujarat and India also.
5. To understand feelings of the farmers about the extension management ability of KVKs, one opinion study can also be conducted.
6. Latest agricultural knowledge and skill are the important factors to make programme coordinators as able and effective managers of KVKs, thus to understand existing level of knowledge and skill about scientific crop production, animal husbandry, dairy, value addition in agriculture, agricultural disaster management, sustainable agriculture, compost farming, integrated pest management, integrated nutrient management, integrated crop management, agricultural Government strategy, WTO, agricultural export, quality production, precise farming, hi-tech farming etc., a study can be conducted.
7. The Computer has become an integral part of daily life, thought it is not being used by many extension personnel. The use of computers and information technology can make programme coordinators of KVK effective and efficient managers, in spite of this, many programme coordinators of KVKs do not prefer to use computers and information communication technologies due to computer anxiety. Thus, a study to know level of computer anxiety

among the programme coordinators can also be undertaken.

8. There are wide scope of internet for agricultural extension, thus a study on the awareness, attitude and views of the programme coordinators about internet can also be undertaken.
9. In the present study, some of the variables have been studied as independent variables. Some others independent variables which can play a role on the extension management ability of the training organizers, so new study can be undertaken by adding those untested and unstudied variables.
10. Some other aspects of programme coordinators such as extension productivity, role expectations, role performance, role-taking empathy etc. may be undertaken.
11. Such study on coordination, research-extension linkage and effectiveness or impact of KVK can be undertaken.
12. Such studies should be repeated after some lapse of time.

References

Alvarez, A. and Arias, C. (2002). Diseconomics of size with fixed management ability. *American J. of Agri. Econ.*, **85** (1): 134-142.

Anonymous (1998). Faculty development in extension management. *University News*, **36** (44): 27.

Anonymous (1998)* Communication excellence in management, *University News*, **35** (50): 51.

Anonymous (1999). Guidelines for the KVK Manager, New Delhi.

Anonymous (1999)*. Productivity. *Agricultural Today*, **2** (3): 58-59.

Anonymous (2000). Green revolution, the need of the hour. *Agricultural News*, **9** (2): 65.

Anonymous (2000)*. Green revolution, the need of the hour. *Agricultural News*, **9** (2): 68.

Anonymous (2001). *News Digest*, NDDB, Anand, Dec. 2001.

Anonymous (2002). Green revolution, the need of the hour. *Agricultural News*, **6** (3): 63.

Anonymous (2005). *Annual report 2004-05*, Department of Agricultural Research and Education, Ministry of Agriculture, Govt. of India, ICAR, New Delhi.

Anonymous (2005)*. *Speech of Mr. Sharad Pawar*, Union Minister of Agriculture, Govt. of India, *National Conference on KVKs – 2005*, NASC Complex, New Delhi, October 27-28.

*Appley, S. A. (1990). Motive strength and leadership style as predictors of managerial effectiveness in selected community colleges. *Dissertation Abst. International*, **40** (9-A): 4812.

Ardeshana, V. R. (1989). Teacher's attitude towards teaching at the College of Agriculture, Junagadh. *M. Sc. (Agri.) Thesis (Unpublished)*. GAU, Sardar Krushinagar.

*Barnard, C. (1938). Social system theory. In: Reddy, P. N., Tripathi, P. C. and Appannaiah, H. R. 1990. Essential of management. Bombay, Himalaya Publishing House, pp. 24-25.

Bhandari, M. C. (1998). Professionalisation of cooperative management: Cooperative perspective. *J. of Coop. Management*, **5**: 15-17.

Bhatia, B. S.; Prem, K. and Chawla, A. S. (1990). Management of rural development (Ed.). New Delhi, Deep and Deep Publication, pp. 1-18.

Bhatia, B. S.; Prem, K. and Chawla, A. S. (1990)*. Management of rural development (Ed.). New Delhi, Deep and Deep Publication, pp. 1-18.

Bhimjiani, A. M. (1978). Study of personality characteristics as related to job performance of agricultural extension officers in Gujarat. *M. Sc. (Agri.) Thesis (Unpub.)*, G. A. U., Navsari.

Bhople, R. S. and Palaspagar, V. W. (1996). Appraisal of managerial ability of farm supervisors, *PKV. Res. J.*, **20** (2): 161-162.

Borkar, M. M. and Chimurkar, B. S. (1996). Diagnostic ability of scientists engaged in T and V programme. *Mah. J. of Ext. Edu.*, **15**: 198-202.

*Brech, E. L. F. (1972). Administration and management in: Reddy, P. N., Tripathi, P. C. and Appannaiah, H. R. (1990). Essential of management. Bombay Himalaya Publishing House-9.

Chandel, R. S. (1974). A handbook Agricultural Statistics, Achal Prakashan Mandir, Kanpur.

Chaudhary, P. K.; Sharma, S. S. and Gour, R. A. (1989). Job performance and job satisfaction of agricultural development officers. *Mah. J. of. Ext. Edu.*, **8**: 245-247.

Chari, A. P. and Nandapurkar, G. G. (1989). A scale to measure managerial ability of farmers. *Mah. J. of. Ext. Edu.*, **6**: 163-198.

Chaya, P. (1974). An investigation into certain psychological characteristics of an effective schoolteacher, *Ph. D. Thesis (Unpub.)*, Kanpur, University, Uttar Pradesh.

Chitra Pathak; Manish Kumar and Kwatra, J. K. (2005). Extension Training for Agricultural Development. *Agri. Ext. Review*, **17**: 17-22.

Daftur, D. N. (1986). Delegation of authority scale. Department of Psychology, M. S. University, Baroda In: Restonjee, D. M. 1989. Second Handbook of Psychological and Social Instruments (Ed.) New Delhi, Concept Publishing Company.

Dakhore, K. M. and Bhilegaonkar, M. G. (1988). Job satisfaction of veterinari extension personnel. *Mah. J. of. Ext. Edu.*, 7: 75-80.

Das, P. (2003). National Agricultural Technology Project (NATP). Indian Council of Agricultural Research. *Proceedings of 26th meeting of the project management committee (PMC) of NATP*, September, 2003.

Dasgupta, A (1969). Indian business management, Dept. of Business Management and Industrial Administration, University of Delhi: 60.

*Davis, R. C. (1951). Management defined In: Aggarwala, D. V. 1995. Principles practice techniques. New Delhi, Deep and Deep Publications, pp. 26.

*Drucker, Peter (1954). The practice of management. New York, Harper and Row Publishers.

*Drucker, Peter and Fayol, H (1954). Administration and Management. In: Reddy, P. N., Tripathi, P. C. and Appanniaiah, H. R. 1990. Essentials of management. Bombay, Himalaya Publishing House, pp. 7-8.

*England, G. W.; Agrawal, N. C. and Trerise, R. E. (1971). Union leaders and Managers: a comparison of value system. Industrial Relation, **10**: 211-214.

*Fayol Henry, (1949). General Industrial management. London SRI ISSAC Pitman and Sons.

Gardebroek, C. L. and Lansink, A. O. (2003). Estimating farm productivity differentials using panel data. *J. of Agri. Econ.*, **54** (3):397-415.

*Gantt, H. L. (1919). Early management approaches In: Reddy, P. N., Tripathi, P. C. and Appannaiah, H. R. 1990. Essential of management. Bombay Himalaya Publishing House, pp. 20.

*Gilbreth, F. (1915). Early management approaches In: Reddy, P. N., Tripathi, P. C. and Appannaiah, H. R. 1990. Essentials of management. Bombay, Himalaya Publishing House, pp. 20.

Gilbreth, F. (1915). Early management approaches In: Reddy, P. N., Tripathi, P. C. and Appannaiah, H. R. 1990. Essentials of management. Bombay, Himalaya Publishing House, pp. 45.

*Gullick, L. (1925). Process management In: Reddy, P. N., Tripathi, P. C. and Appannaiah, H. R. 1990. Essential of management. Bombay. Himalaya Publishing House, pp. 21-22

*Hellriegel, D and Slocum, J. W. Jr. (1975). International Climate: Measures, research and contingencies. *Academy of Management J.*, **17**: 225-280.

Hodge, B. J. and Johnson, H. J. (1970). Management defined In: Aggrawal, D. V. 1995. Management–Principles–Techniques New Delhi Deep and Deep Publications, pp. 25.

Igodan, Charis O. (1987). Job stress and burnout in extension work. *Ind. J. of. Ext. Edu.*, **23** (1 and 2): 1-6.

Islam, M. M. and Alli, M. H. (1991). Impact of training as perceived by the immediate workers. *Ind. J. of Ext. Edu.*, **27** (1 and 2): 81-87.

Jaggi, B. L. (1978). Management leadership styles in Indian work organizations. *Indian Manager*, **9** (2): 139-156.

Jain, Uma R. (1981). Life styles of Indian managers: A study of district collectors. *Ph. D. thesis (Unpub.)*, Indian Institute of Management, Ahmedabad.

Jondhale, S. G.; Ingle, L. A. and Pathak, U. N. (2000). Impact of Krishi Vigyan Kendra Training on adoption of improved practices of summer groundnut. *Mah. J. of. Ext. Edu.*, **19**: 109-112.

Kalra, B. S. and Mishra A. K. (2005). Village level framework for watershed management. *Agri. Ext. Review*, **17**: 23-26.

Kareem, K. A. and Jayaramaiah, K. M. (1997). Problems perceived by the staff in the planning and implementation of integrated Rural Development Programme. *Madras Agric. J.*, **84** (7): 354-356.

Kerlinger, F. N. (1976). Foundation of Behavioural Research, New Delhi, Surjee Publication, pp. 198-204.

Khare, H. C.; Jaiswal, D. K. and Mishra, P. K. (1987). Role performance of SMSs in T and V system. *Maha. J. Ext. Edu.*, **6**: 199-200.

Kholn, M. L. and Schooler, C. (1973). Occupational experience and psychological functioning. An assessment of reciprocal effects. *American Sociological Review*, **38:** 97-118.

Koontz, H. D. and O'Donnell, C. J. (1972). Principles of management An analysis of managerial functions Tokyo, McGram Hall, Kogākusha, pp. 5-20.

Koontz, H. D.; O'Dennell, C. J., and Weihrich, H. (1980). Management Auckland, McGraw Hill International Book Company, pp. 1-15.

Laharia, S. N. (1978). A study of personal and organizations variables influencing the productivity of agricultural scientists. *Ph. D. (Agri.) Thesis (Unpub.)*, Haryana Agricultural University, Hissar.

Lee, Jung Hwan and Kim Song Soo (2004). The road ahead for Korean agriculture soft landing on a new plateau in the WTO. Seoul, Korean Rural Economic Institute.

Lodalh, T. M. and Kejner, M. (1965). The definition and measurement of job involvement. *J. Appl. Psychology*, **49**: 24-33.

Malik, C. P. (1988). A study on job satisfaction of the teachers of Gujarat Agricultural University, Anand. *M. Sc. (Agri.) Thesis (Unpub.)*, G. A.U., Sardar Krushinagar.

Malik C. P. and Patel, H. L. (1990). Factor related with job satisfaction in University. *Mah. J. of. Ext. Edu.*, **9**: 42-45.

Mangala, Rai (2005). *Annual Report of ICAR*, Department of Agriculture (DARE), Ministry of Agriculture, GOI, New Delhi.

Mangala, Rai (2006). *An inaugural speech at 76th Annual general meeting of the ICAR society*, ICAR, New Delhi.

*Mayo, Elton. (1932). Human relation movement. In: Reddy, P. N., Tripathi, P. C. and Appannaiah, H. R. 1990. Essentials of management. Bombay, Himalaya Publishing House, pp. 22-23.

*McGreger, D. (1967). The professional manager. New York, McGraw Hill, pp. 5-7.

McGreger, D. (1971). The human side of enterprise. In: Rechards, M. D. and Neilander, W. A. (Eds.) Bombay, D. B. Taraporewala Sons and Co. Ltd.

Mee, J. F. (1963). Management defined In: Aggarwal, D. V. 1995. Management: Principles-Practices-techniques. New Delhi, Deep and Deep Publications, pp. 24.

Mehta, S. L. (1998). Convocation address in 23rd annual convocation at Gujarat Agricultural University, Sardar Krushinagar.

Monappa, A. and Saiyaddain, M. S. (1997). Personal management. New Delhi, McGraw Hill, pp. 327.

Morell, R. W. (1969). Management: Ends and Means. Sanfransisco, Chandler Pub. Co. pp. 134-143.

Murthy, R. and Somasundaram, D. (1989). Factors associated with the role performance of Village Extension Officers. *Maha. J. Ext. Edu.*, **8**: 28-34.

Nathall, P. L. (2001). Managerial ability a review of its basis and potential improvement using psychological concepts. *J. of Agric. Econ.*, **24** (3): 247-262.

Nikhade, D. M. and Kitey, P. V. (1984). Role performance of village level workers in agricultural development programme. *Mah. J. of. Ext. Edu.*, **3**: 111-112.

*Niles, M. C. (1956). The Essence of Management Bombay, Orient Longman, pp. 20-24.

Padheria, M. M. (1989). Scientific productivity of academic staff from Marathwada Agricultural University, Parbhani: a system analysis. *Ph. D. (Agri.) thesis*, Marathwada Agricultural University, Maharashtra.

Padheria, M. M.; Soni, M. C. and Kulkarni, R. R. (1992). Prediction of scientific productivity of academicians. *Guj. J. Ext. Edu.*, **1**: 1-4.

Pal, S. and Singh, Alka (1997). Agricultural Research and Extension in India: Institutional structure and investments. Policy Paper. No. 7, NCAR and PR. New Delhi. pp. 15-18 and 59-63.

Parmar, S. D. (2005). Publication behaviour of the Agricultural Scientists of Anand Agricultural University, Anand. *M. Sc. (Agri.) Thesis (Unpub.)*, A.A.U., Anand

Patel, C. S. (1990). Important qualities of agricultural college teacher as perceived by students and teachers of agricultural college. *Guj. J. of Ext. Edu.*, **1**:93.

Patel, D. J. (1990). A study on teaching effectiveness of teachers working in Gujarat Agriculture University with special reference to faculty of agriculture. *Ph. D. thesis (Unpub.)*, G.A.U., Anand.

Patel, D. P. (1992). Effectiveness of mass media on psychological attributes of agricultural teachers in Kheda district of Gujarat state. *M. Sc. (Agri.) Thesis (Unpub.)*, G.A.U., Sardar Krushinagar.

Patel, A. A. (1993). Study on Research management ability of Research Scientist working as Head of Department/Station/ Project/Scheme in G.A.U., *Ph. D. Thesis (Unpub.)*, G.A.U., Anand

Patel, A. A.; Patel, M. R. and Desai, C. P. (1998)*. Research management ability of the heads of research departments in Gujarat Agricultural University. *G AU. Res. J.*, **23** (1): 56-59.

Patel, A. A.; Patel, M. R. and Desai, C. P. (1999). Job related and socio-psychological factors affecting research management ability of the heads in Gujarat Agricultural University. *G AU. Res. J.*, **23** (2):77-80.

Patel, G. J. (2001). A study on extension management efficiency of Training Organizer in KVK. *Guj. J. of Ext. Edu.*, Dec.-2001. Vol. 14: 18.

Patel, G. J. (2001). A study on extension management ability of Training organizers on KVKs. *Ph. D. Thesis (Unpub.)*, G.A.U., Sardar Krushinagar.

Patel, G. J. and Patel, A. A. (2003). Extension management power of TOs of KVKs in Gujarat state. *Guj. J. of Ext. Edu.*, Dec. 2003. Vol. **18**: 57-59.

Patel, G. J.; Soni, M. C. and Thakkar, K. A. (2004). A scale to measure extension management ability of training organizer of KVKs. *Guj. J. Ext. Edu.*, **15**: 29-32.

Pawar, Sharad (2006). Presidential speech of Agricultural Minister at 76th Annual general meeting of the ICAR society at the NASC complex in New Delhi.

*Pearsons, T. (1954). Essays in Sociological theory Glencoe. The free press, pp. 275-286.

Popat, M. N. (1991). A study on productivity of SMSs working under T and V system in Gujarat State. *Ph. D. Thesis (Unpub.)*, G.A.U., Anand.

Prasad, C. (1991). International conference on extension strategy for minimizing risk in rainfed agriculture. *Ind. J. Ext. Edu.*, **27** (1 and 2): 1-12.

Rani G. Janshi (1985). Scientific productivity of agricultural scientists: An activity analysis approach. *Ph. D. (Agri.) Thesis (Unpub.)*, Andhra Pradesh Agricultural University, Hyderabad.

Rao, R. and Mohan, R. (1986). A study on job satisfaction of teachers of Andhra Pradesh Agricultural University and factors associated with job satisfaction. *M. Sc. (Agri.) Thesis (Unpub.)*, A.P.A.U., Hyderabad.

Rao, R. R. (1986). A study of job satisfaction of Andhra Pradesh Agricultural University and associated with job satisfaction. *M. Sc. (Agri.) Thesis. (Unpub.)*, APAU, Hyderabad.

Reddy, M. S. and Jayaramaiah, K. M. (1988). Relationship between socio-psychological characteristics and the job effectiveness of VEWs. *Ind. J. of. Ext. Edu.*, **24** (3 and 4): 88-89.

Reddy, M. S. and Jayaramaiah, K. M. (1990). Relationship of selected variables on productivity of village extension officers. *Ind. J. Ext. Edu.*, **26** (1 and 2): 78-81.

Reddy, R. and Rao, B. V. (1998). Role performance and job satisfaction of village extension officers working with intensive agricultural extension programme (T and V system). In: Desai, G. R. and Reddy, M. R. (1986). Ed. Studies on T and V Extension in India, NIRD, hyderabad, **2**: 80-81.

Reddy, M. N. and Sinha, B. P. (1990). Factors influencing the role conflict among the scientists in an agricultural university system. *Mah. J. of. Ext. Edu.*, **9** (1 and 4): 1-4.

*Rogers, E. M. (1962). Diffusion of innovation, New York, The free press of Glencoe, pp. 214-217.

Sah, A. K. (1999). Challenges before agricultural extension into 21st century. *Agril. Ext. Review*, **11** (5): 1-8.

Sakaria, V. B. (1988). Role expectation, role preference, role performance and training needs of SMSs working under T and V system in Gujarat State. *Ph. D. (Agri.) Thesis (Unpub.)*, G.A.U., Junagadh.

Samantha, D. K. (1985). Perception of work environment by agricultural research service scientists of ICAR institute. *Ind. J. Ext. Edu.*, **21** (3 and 4): 53-61.

Samanta, D. K. (1991). Agricultural extension in changing world perspective. New Delhi, Uppal Publishing House, pp. 1-20 and 121-143.

Sawant, A. G. (2004). Role of informatics and bio-informatics in extension education in India. *Asian J. of Ext. Edu.*, **23** (1): 22-41.

Scanlan, B. K. (1973). Principles of management and organizational behaviors Canada, John Wiley and Sons Inc.

Sharma, M. L.; Sharma, P. N.; Jaiswal, P. K. and Sengar, R. S. (1988). Role expectation and role performance of rural agricultural extension officers of T and V system in M. P. *Indian J. of. Ext. Edu.*, **24** (3 and 4): 75-78.

Shah, U. B. (2006). A study on level of Internet exposure of Teachers of Anand Agricultural University, Anand. *Ph. D. Thesis (Unpub.)*, A.A.U., Anand, Gujarat.

Shivram, M. and Dalal, Y. C. (1994). Lekh Lakhvani Kala. (Art of writing: Article) Pravin Prakashan, Rajkot, Gujarat, p. 19.

Singh, J. P. (2000). WTO and India's agricultural export perspectives. *Manage Ext. Res. Review*, **1** (2): 1-11.

Singh, R. (2001). Rural infrastructure, agricultural development and poverty in India. An inter state study. *J. of Rural Development*, **23** (1): 31-57.

Singh, S. D.; Joshi, K. J. and Sharma, N. K. (2004). Mass media exposure among the horticultural development officers in Haryana. *Rural India*, **67** (5): 90-94.

Singh, Baldeo and Padaria, R. N. (2004). Frontiers of scientific quests today in agricultural extension. *Asian J. of Ext. Edu.*, **22** (1): 104-121.

Singhi, P. M.; Wadwalkar, S. and Kaur, G. (1982). Management of Agricultural Extension. Monogram No. 96; Indian Institute of Management, Ahmedabad, pp. 1-12.

Sendecor, G. W. and Cochran, W. G. (1967). Statistical methods VI. Ed. Oxford and IBI Publishing Co. Pvt. Ltd. New Delhi.

Soni, N. V. (1990). Content analysis of Krushi Go Vidya and Krushi Jivan farm magazines: Special problem. *M. Sc. (Agri.) Thesis*, G.A.U., Junagadh.

Srivastava, V. L. (1981). Contours of management. New Delhi classical Publishers, pp. 1-8.

Stagner, R. P. (1956). Perception: applied aspects. In the psychology of industrial conflict. John. Wiley and Sons, New York.

Sudesh, Rani (2005). A study on gain in knowledge and attitude of trained women towards fruit and vegetables preservation training of Vadodra district. *M. Sc. (Agri.) Thesis (Unpub.)*, A.A.U., Anand.

Sukhadia, A. G. (2005). Author's motivation towards writing articles in Krushi Jivan magazine. *M. Sc. (Agri.) Thesis (Unpub.)*, A.A.U., Anand.

Sulaiman, V. R. and Hall, A. (2004). Towards extension – plus: Opportunities and Research, Policy Brief 17 New Delhi.

Swanson, B. E. (2004). Extension strategies for poverty alleviation in a global economy. Paper presented at the international seminar on "Best practices in poverty alleviation through education for the new century". Suwom, South Korea.

Talukdar, R. K. (1984). Productivity of agricultural development officers in Haryana: A factor analysis study. *Ph. D. Thesis (Unpub.)*, H.A.U., Hissar.

*Taylor, F. W. (1915). Early management approaches In: Reddy, P. N., Tripathi, P. C. and Appannaiah, H. R. 1990. Essentials of management. Bombay, Himalaya Publishing House, pp. 19-20.

Terry, G. R. (1968). Meaning of Management. In: Reddy, P. N., Tripathi, P. C. and Appannaiah, H. R. 1990. Essentials of management. Bombay, Himalaya Publishing House, pp. 20.

Thakkar, D. (1986). A study on the adoption of recommended summer groundnut technology by the farmers of Anand talukas of Kheda district of Gujarat state. *M. Sc. (Agri.) Thesis (Unpub.)*, G.A.U., Anand.

Tolmay, V. L. and Hernandez, P. (2001). Resistance to biotic and abiotic stress in the Triticeae, *proceedings of fourth international Triticeae Symposium*, Cordoba, **135** (2 and 3): 239-242.

Trip, G.; Thijssen, G. L.; Renkeema, J. A. and Huirne, R. B. M. (2003). Measuring managerial efficiency: the case .of green house growers *J. Agric. Econ.*, **27** (2): 175-181.

Tripathi, B. D. (2000). Agricultural Extension personnel in new millennium: A prospective view. *Manage Ext. Res. Review*, **1** (2): 12-23.

Unger, P. W.; Skidmore, E. L. and Carter, M. R. (1995). Conservation tillage in the southern United States Great Plains, conservation tillage in temperate agro-ecosystems, U. S. A.: 329-356.

Urwick, L. and Gullick, L (1925). Process Management In: Reddy, P. N., Tripathi, P. C and Appannaiah, H. R. 1990. Essential of management, Bombay, Himalaya Publishing House, pp. 21-22.

Van de Ban, A. W. (2004). Changes in society require changes in agricultural extension: some observations, *Asian J. of Ext. Edu.*, **23** (1): 1-17.

Van den Ban, A. W. and Hawkins, H. S. (1996). Agricultural extension. London, Blackwell Science Ltd., pp. 1-6 and 233-266.

Vance, S. (1956). Meaning of Management. In: Reddy, P. N., Tripathi, P. C. and Appannaiah, H. R. 1990. Essentials of management. Bombay, Himalaya Publishing House, p. 3.

Veerbhadraiah, V. (1983). Time management, job involvement and job performance of extension supervisors. *Thesis Abstract*, **9** (2): 124.

Veerbhadraiah, V. and Jalihal, K. A. (1983). A study on time utilization of deputy directors and assistant director of agriculture on different activities. *Mysore J. Agric. Sci.*, **9** (2): 342-347.

Venkateshwarlu, N. (1988). Foundation course of agricultural research project management course. No. 28, Lecture Notes, NAARM, Hyderabad.

Verma, O. S. (1997). Management of Extension system: Authority, responsibility and accountability. *J. of Ext. System*, **13** (1 and 2): 20-35.

Vinayangam, S. and Samanta, R. K. (2004). Assessment of work environment among trainers in KVKs. *Asian J. of Ext. Edu.*, **23** (1): 129-134.

Vroom, V. H. (1964). Work and motivation New York, John Wiley and Sons Inc.

Wase, R. B. (2001). Knowledge and adoption of farmers about jayanti Chilli cultivation *M. Sc. (Agri.) Thesis (Unpub.),* P.D.K.V., Akola, Maharashtra.

Woodworth, R. S. (1979). Assessment of personality (Scale): Personality. New Delhi, Oxford.

Wu-Sx; Deva Doss, S.; S and Lu, Y. C. (2003). Estimation and decomposition of tech. efficiency for sugar beet farms. Applied Economics, U.S.A., 35 (4): 471-484.

* Original not seen

Appendices

Appendix I

Interview Schedule

Extension Management Ability of the Programme Coordiantors of KVK

Name of the respondents : ____________________

Name of KVK : ____________________

Whether KVK belongs to SAU/ICAR/NGO/Deemed University

Part – I indcpendent variables

(A) Personal Information

1. Age of the respondents in completed years: __________
2. Educational qualification: __________
3. What do you feel about your health (Very good/Good/Average/Poor/Very poor)
4. Type of family (a) Joint ______ (b) Nuclear ______
5. Total no. of family member ______
6. Native place (a) Rural ______ (b) Urban ______

(B) Job Related Information

1. Total experience as training organizers______ years
2. Total working experience in different cadres (KVK) ______ years

3. Total No. of training received during service period

Short term training (up to 7 days) Nos. _______

Medium term training (8 to 15 days) Nos. _______

Long term training (above 15 days) Nos. _______

Total Nos. _______

Part – II Organizational information

(A) Span of Control

Give total no. of staff working under your guidance

Cadre	*No.*
Technical	
Administrative	
Auxiliary	
Total	

(B) Staff Facility

what do you feel about staff facility provided to you in different cadres to carryout extension activities of KVK (ü) your response

Sl.No.	*Category*	*More than Required (Excess) (2)*	*As per Requirement (Sufficient) (3)*	*Less than Required (Insufficient) (1)*
1.	Technical			
2.	Administrative			
3.	Auxiliary			

(C) Organizational Climate

Please indicate your response by putting (✓) against each statement which explains the level of organizational climate of your KVK.

Sl.No.	*Statements*	*Response*			
		Highly	*Quite*	*Some what*	*Not at all*
		(4)	*(3)*	*(3)*	*(1)*
1.	Staff is generally interested in welfare and happiness of each other				
2.	Staff is clear about the goal and objectives of KVK				
3.	Differences and disagreement are not avoided but sorted out				
4.	Staff is satisfied here with their job				
5.	Staff feels responsible for effective functioning of KVK				
6.	Staff has confidence and trust in each other				
7.	Information is shared by the staff freely				
8.	Staff is generally gives ideas and suggestions to each other				
9.	There is an atmosphere of warmth and supporting in the KVK				
10.	Staff is generally bothered if some one faces difficulty				
11.	There are opportunity to use and develop capacities of staff				

Contd...

Contd...

Sl.No.	Statements	Response			
		Highly (4)	Quite (3)	Some what (3)	Not at all (1)
12.	Staff takes initiative to improve the work of KVK				
13.	There is a good prevailing team spirit				
14.	Good work is always appreciated				
15.	Staff is concerned with effective performance and use innovative was rather than referring the rules				

(D) Organizational Facilities

Please indicate which of the following facilities available with your KVK.

Sl.No.	Facilities	Available	Easily available from other	Available with some difficulty	Not Available
A.	**Training facilities**				
1.	Audio-visual				
2.	Training hall / lab.				
3.	Training material				
4.	Reading material				
5.	Photographic facility				
6.	Videography				

Contd...

Contd...

Sl.No.	*Facilities*	*Available*	*Easily available from other*	*Available with some difficulty*	*Not Available*
7.	Hostel for trainers				
8.	A-V equipment				
B.	**Other facilities**				
1.	Office building				
2.	Office furniture				
3.	Telephone				
4.	Fax				
5.	Xerox				
6.	Computer				
7.	Internet connection				
8.	Cyclostyle machine				
9.	Typewriter				
10.	Vehicle for extension work				
11.	Others, if any				

(E) Interpersonal Communication

How frequently do you discuss about your activities / problems with the following persons? Tick mark (✓) in the appropriate column.

Sl.No.	*Persons*	*Response*				
		Very Frequently (5)	*Freque-ntly (4)*	*Some-times (3)*	*Seldom (2)*	*Never (1)*
1.	Head of your organization					
2.	Officers of ICAR					
3.	Officers of SAU					
4.	Officers of state Dept. of agriculture					
5.	Officers of line department at district level					
6.	Technical/ research staff					
7.	Administrative staff					
8.	Others, if any					

Part – III Socio-Psychological variables

1) Personality of Training Organizers

The following questions are relating to your personality. Please answer each statement by marking (✓).

Sl.No.	*Questions*	*Answer*	
		Yes	*No*
1.	Do you like to talk before a group of people		
2.	Are you some what suspicious of other people motives		
3.	Do you always try to make others agree with you?		
4.	Do you suffer from the feeling of inferiority?		
5.	Do you make friends easily?		
6.	Are you feeling easily hurts?		
7.	Are you at home among strangers?		
8.	Do you worry over what people think of you?		
9.	Do you like to take the lead in a social gathering?		

2) Attitude towards Extension Work

You are requested to check (✓) appropriate response category against each statement expressing your attitude towards extension work.

Sl.No.	Persons	Attitude towards Extension Work SA	A	U	DA	SD
1.	Honestly speaking, I hate extension					
2.	Extension work is the need of the hour.					
3.	Money spent on extension work is sheer wastage.					
4.	I really enjoy meeting farmers and doing extension work					
5.	I am trapped in extension profession					
6.	The future of the nation lies only in extension work					
7.	Extension work is a thankless job.					
8.	Extension work is a good medium of serving humanity					
9.	Extension workers are considered as second class citizen.					
10.	I have poor status in the eyes of relatives only because I am working as training organizer.					

SA = Strongly agree, A = Agree, U = Undecided, DA = Disagree, SD= Strongly disagree

3) Perceived of Workload

What do you perceive about your workload? Please (✓) against the options given below.

Light workload (1) ( )

Sufficient workload (2) ( )

Heavy workload (3) ( )

4) Job Involvement

The following statements regarding orientation to job are given, please indicate your response about your job involvement by putting (✓) against each statement.

Sl.No.	*Statements*	*Response Categories*				
		SA	*A*	*U*	*DA*	*SD*
1.	I have other activities more important than my work.					
2.	To me work is only a small part of my life					
3.	Quite often I feel like staying at home instead of going for work.					
4.	I used to care more about my work but now other things are important to me.					
5.	We can measure a person pretty well by how a good job he does.					
6.	I shall stay overtime to finish a job, even if I am not paid for it.					
7.	I live, eat and breath for my job.					
8.	Sometimes I would like to kick my self for the mistakes I make in my work					

Contd...

Contd...

Sl.No.	*Statements*	*Response Categories*				
		SA	*A*	*U*	*DA*	*SD*
9.	For me nothing at work really go off quickly					
10.	I used to be more ambitious about my work than I am now.					
11.	I usually go for work a little early to get the things ready.					
12.	I am really a perfectionist about my work.					
13.	I avoid taking extra duties and responsibilities in my work.					
14.	Most things in life are more important than work.					
15	I feel depressed when I fail at something connected with my work.					
16.	Sometimes I keep myself awake at night thinking ahead to the next days work					
17	The major satisfaction in my life comes from my work					
18.	I would keep working even if I do not get the money.					
19.	I am very much involved personally in my work.					
20.	The most important things that happen to me involves my work.					

SA = Strongly agree, A = Agree, U = Undecided, DA = Disagree, SD= Strongly disagree

5) Job Stress

Given below are some situations, which arise in your life. These situations process problems, which disturb the balances of life and threaten the person resulting in loss of face, prestige or security in the job life. You are requested to (✓) appropriate response category against each statement expressing your agreement or disagreement with statement.

Sl.No.	*Statements*	*Response Categories*				
		SA	*A*	*U*	*DA*	*SD*
1.	My assignments are of monotonous nature					
2.	I get less salary in comparisons to the quantum of my work.					
3.	My decision and instructions concerning work among staff are properly followed.					
4.	I do my work under tense circumstances.					
5.	Owing to excessive workload, I have to manage with insufficient staff resources.					
6.	I get ample opportunity to utilize my abilities and experiences independently.					
7.	I have to do some work unwillingly owing to certain group/political pressure.					

Contd...

Contd...

Sl.No.	*Statements*	*SA*	*A*	*U*	*DA*	*SD*
8.	Some of my colleagues and sub-ordinates become happy on my unsuccessful events.					
9.	My college does cooperate with me voluntary in solving work problems.					
10.	I have to dispose my work hurriedly due to excessive workload.					
11	I am unable o perform my duties smoothly due to short of authorities.					
12.	Working conditions are satisfactory level for welfare and convenience of staff.					
13.	I often feel that this job has made my life burdens one.					
14.	Being to busy with official work, I am not able to devote sufficient time to my domestic and personal problem.					
15.	The mutual cooperation and team spirit among the staff are existing here.					
16.	I have to do such work as out to be done by others.					
17.	I am unable to carry out the work to my satisfaction due to excessive load of work and lack of time.					

Part - IV

1) Constraints Faced by the Training Organizers

Being a head of the KVK, you may experience Management/ Technical/Administrative constraints which you find obstacles in the effective functioning of the Kendra. Please tick against your (✓) response.

Sl.No.	Constraints	Response Categories		
		MF (3)	F (2)	NA(1)
1.	Many posts of the subordinate staff lying vacant			
2.	Majority of the subordinate staff not fulfilling the required educational qualification			
3.	Frequent transfer of the staff from KVK to other department.			
4.	Funds for the extension activities/ FLDs not released in time by the council.			
5.	Most of the staff is engaged with organizational work (other than KVK activities)			
6.	No promotion channel for the KVK staff.			
7.	Inadequate infrastructure facilities such as building, fencing, borewell, vehicle etc.			

Contd...

Contd...

Sl.No.	Constraints	Response Categories MF (3)	F (2)	NA(1)
8.	Unnecessary involvement of organizational head in taking decision regarding planning, execution and implementations of the KVK activities.			
9.	Unfair contact of the organizational head with subordinate staff.			
10.	Intermittent flow of the funds from organization to KVK head for carrying out mandated activities.			
11.	Pressure from the local politician to fetch more benefits from KVK schemes to their own jurisdiction.			
12	Organizational rules not allow the KVK staff to work freely according to the guidelines of council.			
13	If any, (you can add)			

MF = Most frequently, F = Frequently, NA = Not at all

1) Please give your suggestions to overcome the constraints.

1. ____________________
2. ____________________
3. ____________________
4. ____________________
5. ____________________

Part - V

Information Pertaining to Extension Management Ability of Training Organizers

You have to perform many functions related to management of extension activities. Following are the statements regarding management functions. Please indicate your perception against each statement putting (✓) marks.

1) Planning

How often do you plan about different functions?

Sl.No.	*Sub Indicators with Weightages*	*Always* (3)	*Some times* (2)	*Never* (1)
1.	To determine the objectives			
2.	To ascertain training needs			
3.	To plan for transfer of technologies			
4.	To plan for utilization of funds			
5.	To plan for extension facilities			
6.	To prepare annual plan of work			
7.	To plan for monitoring work			
8.	To plan for evaluation work			

(2) Organizing

How often do you organize different functions?

Sl.No.	*Sub Indicators with Weightages*	*Always* (3)	*Some times* (2)	*Never* (1)
1.	To call regular meeting			
2.	To provide work opportunity			
3.	To provide work according to specializations			
4.	To invite suggestions from subordinates			
5.	To divide the activity into groups			
6.	To delegate authority according to responsibility			

(3) Directing

How often do you direct subordinates about different functions?

Sl.No.	*Sub Indicators with Weightages*	*Always* (3)	*Some times* (2)	*Never* (1)
1.	To give clear instruction of work			
2.	To give due credit to a person			
3.	To appreciate good work done			
4.	To give continuous information			

(4) Communitating

How often do you involve to communicate different functions?

Sl.No.	*Sub Indicators with Weightages*	*Always* (3)	*Some times* (2)	*Never* (1)
1.	To give timely instructions			
2.	To listen to subordinates			
3.	To select proper medium to instructions			
4.	To receive feed back			

(5) Controlling

How often do you control different functions ?

Sl.No.	Sub Indicators with Weightages	Always (3)	Some times (2)	Never (1)
1.	To monitor the work			
2.	To visit accurately and properly			
3.	To determine standard for work			
4.	To evaluate the work			
5.	To measure performance of staff			

(6) Human relation

How often do you make personal touch with persons to carry out functions ?

Sl.No.	Sub Indicators with Weightages	Always (3)	Some times (2)	Never (1)
1.	To make personal contacts			
2.	To deal fairly			
3.	To take interest in development of juniors			
4.	To adopt participative approach			
5.	To understand others behaviour			
6.	To provide co-operation			

(7) Leading

How often do you take lead in different functions ?

Sl.No.	Sub Indicators with Weightages	Always (3)	Some times (2)	Never (1)
1.	To promote team work			
2.	To take initiative task			
3.	To apply technical competency			
4.	To have a sense of responsibility			
5.	To inspire subordinates			

(8) Supervising

How often do you supervise different functions ?

Sl.No.	Sub Indicators with Weightages	Always (3)	Some times (2)	Never (1)
1.	To provide wide counseling			
2.	To focus on quality of work			
3.	To give suggestions			

(9) Coordinating

How often do you co-ordinate different functions ?

Sl.No.	*Sub Indicators with Weightages*	*Always* (3)	*Some times* (2)	*Never* (1)
1.	To establish linkages			
2.	To integrate the extension work			
3.	To establish mutual trust			
4.	To get co-operation from outside			

(10) Decision making

How often do you take proper decision about different functions ?

Sl.No.	*Sub Indicators with Weightages*	*Always* (3)	*Some times* (2)	*Never* (1)
1.	To take decision at proper time			
2.	To find out facts for decision			
3.	To adopt participative decision			
4.	To implement decision			
5.	To take decision free from biases			

Appendix II

Instruments Used for Measurement of Variables

The variables studied and instruments used for their measurement are summarized as follows:

Sl.No.	*Variables*	*Instrument Used for Measurement*
(A)	**Dependent Variables**	
1.	Extension management ability	Used scale developed by Patel (2001)
(B)	**Independent Variables**	
1.	**Age**	Structured schedule on chronological age-scoring
2.	Educational qualification	Structured schedule on academic degree-scoring
3.	Health	Scale used as Popat (1991)
4.	Type of family	Structured schedule on type of family-scoring
5.	Size of family	Structured schedule on family members scoring
6.	Native place	Structured schedule on native place - scoring

Contd...

Contd...

Sl.No.	Variables	Instrument Used for Measurement
7.	Experience as a head of KVK	Structured schedule on exp. as head-scoring
8.	Total experience in KVK	Structured schedule on total exp. in KVK-scoring
9.	Training received	Structured schedule in total training received-scoring
10.	Span of control	Structured schedule on span of control total staff-scoring
11.	Availability of technical staff	Structured schedule on technical staff available-scoring
12.	Organizational climate	Jain Uma (1981)
13.	Organizational facilities	Structured schedule on organizational facilities-scoring
14.	Inter personal communication	Structured questionnaire on inter personal communication-scoring
15.	Personality of Training Organizer	Scale developed by Woodworth (1979)
16.	Attitude towards extension work	Scale developed by Daftur (1986)
17.	Perceived work	Structured schedule on work load-scoring
18.	Job involvement	Scale developed by Lodalh and Kejner (1965)
19.	Job stress	Structured questionnaire on job stress-scoring

Plate 17 : Exposure tour organized by KVKs

Plate 18 : Agricultural exhibitions organized by KVK

Plate 19 : Method demonstrations at KVK instructional farm

Appendix III

Feathers of ICAR Spread Over the Country

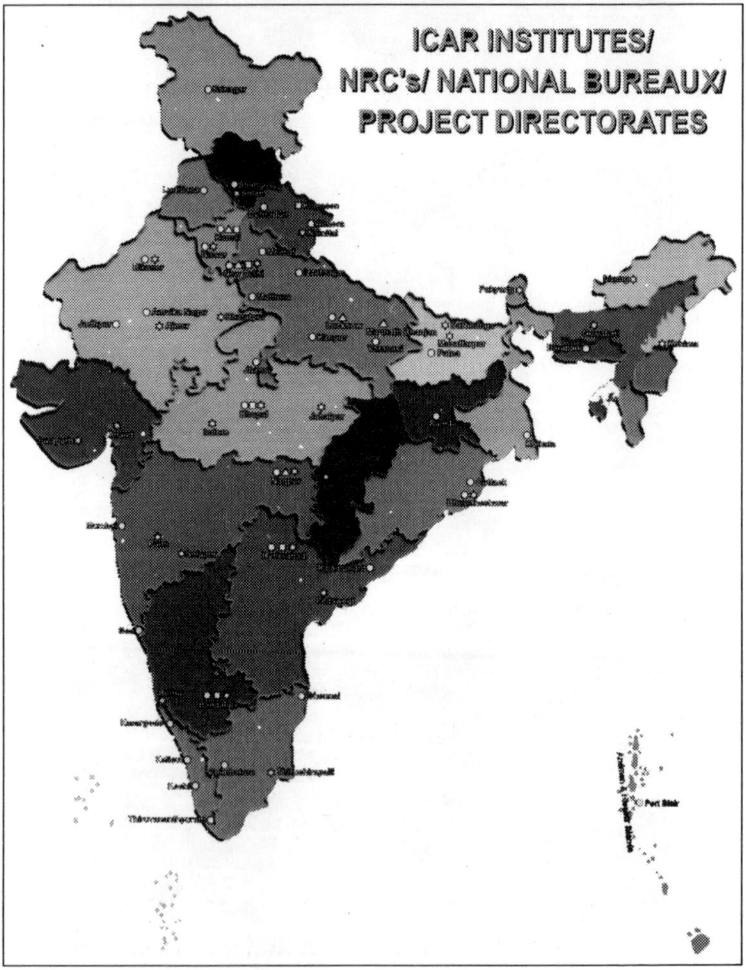

Index